Biochemical Diagnosis of the Elderly

Biochemical Diagnosis of the Elderly

H. M. HODKINSON
M.A., D.M., F.R.C.P.
Consultant Physician in Geriatric Medicine,
Northwick Park Hospital, Harrow

LONDON
CHAPMAN AND HALL

First published 1977
by Chapman and Hall Ltd
11 New Fetter Lane, London EC4P 4EE

Typeset by Santype International Limited, Salisbury
and printed in Great Britain
at the University Printing House, Cambridge

ISBN 0 412 13720 8

Contents

Preface

In the past, elderly patients have often not been investigated in the same depth as younger patients. Correspondingly, until very recent years, little was known either of the difficulties or of the potential benefits of more thorough biochemical laboratory investigation of elderly patients. Until recently indeed, there was almost total ignorance even of such absolutely fundamental information as the normal ranges appropriate to the elderly for biochemical tests. Excellent studies of the age and sex differences in biochemical values, for example that of Roberts (1967), typically stopped short at the age of 65, the use of blood donor populations often being the underlying reason for this.

In recent years, however, thanks particularly to the increasing contribution of autoanalysis, biochemical tests have become far more readily available to elderly patients. Economical use of multi-channel autoanalysers has indeed tended to impose a 'hospital profile' of tests on the clinician whatever his wishes. This is because it is now usually cheaper, quicker, more convenient and more reliable to perform a whole standardised group of tests using a multi-channel autoanalyser than to carry out a smaller number of individually requested single tests. In many clinical disciplines this new development of laboratory profiles is to some extent in conflict with traditional medical attitudes. Classically, laboratory tests have been used in a 'discretionary' manner – that is, for the clinician to base their use on 'his provisional diagnosis or proposals for treatment after examination of the patient' (Holland and Whitehead, 1974). This preference for the discretionary use of tests carries a hint of intellectual snobbery with an implied suggestion that the clinical diagnostic art is in some way debased by the non-discretionary use of tests!

However, in geriatric medicine the increased availability of tests has been an undoubted benefit and has stimulated the rapid development of what had been a seriously neglected field of biochemical investigation. Whilst very little had previously been published, the past few years have seen an ever growing volume of work.

This book attempts to bring together some of the key findings in the developing field of investigation of the elderly. It will particularly draw upon the experience of my colleagues and myself in the geriatric department of Northwick Park Hospital since its opening in 1970, where we have used a reasonably comprehensive admission profile of biochemical tests as a routine. Some aspects of this experience have been published but other general references to 'Northwick Park Hospital' in the following text refer to unpublished data acquired in the department as a result of this admission profile policy. In view of these many references, Appendix II details the main characteristics of the inpatient population of the department.

This is intended to be a practical book on biochemical investigation of the elderly. Whilst heavily based on my own experience in geriatric medicine it is written with the intention that it should be relevant not merely to those practising geriatric medicine but equally to the many other doctors who also treat large numbers of elderly patients whether in hospital or in general practice. It deals with the elderly in general, that is, all those over 65 and not just the very elderly with whom the geriatric physician is particularly concerned.

The differences between elderly and younger patients are relative rather than absolute but principally relate to three special features. Multiple disease is the rule rather than the exception in old age, illness is more often severe and presentation of illness is frequently atypical or non-specific. In consequence, interpretation of biochemical tests is typically more difficult because of the complex effects of multiple disease, disturbed homoeostasis and multiple drug regimes which are more common in ill elderly patients. Furthermore, atypical and non-specific presentations of such treatable conditions as osteomalacia, hypothyroidism or thyrotoxicosis in elderly patients mean that profile tests need to be interpreted in the manner of screening tests and not just in the traditional discretionary way if we are to derive full benefits for the patient and this is an unfamiliar approach for most clinicians.

These extra problems of interpretation are a challenge to the clinician who wishes to derive the maximum amount of useful information from the tests he commonly employs. In addressing myself to these difficulties, I have preferred to examine the interpretation of tests commonly employed more thoroughly rather than to cover a larger number of tests more superficially and without the benefits of sufficient personal experience of the test.

Clinicians in general have difficulty in coping with the sheer volume of profile results (Whitehead and Wootton, 1974) and consequently they often fail to act on many significant abnormalities (Bates and Yellin, 1972). In investigating the elderly the problems are exaggerated by the special difficulties already considered. The clinician is in danger of drowning in a veritable sea of 'abnormal' results which confuse rather than inform. The use of appropriate statistical 'tools' can, however, be of inestimable value in getting the maximum of useful information out of biochemical profiles in elderly patients.

Thus, although this book is intended for the practising clinician dealing with the elderly and is meant to be practical rather than academic or theoretical, statistical considerations will be discussed in some detail. The reader will be assumed to have no special knowledge of statistics and average mathematical ability. The very word statistics tends to strike dread into the clinician's heart but, in the context of this book, I hope that the reader will quickly appreciate that the various statistical tools are practical, common sense aids to interpretation, not mathematical black magic!

London, 1977 Malcolm Hodkinson

Normal ranges for some important biochemical tests in the healthy elderly

Test	95% range	Author
Serum albumin	33–49 g/litre	Leask *et al.* (1973)
Serum globulin	20–41 g/litre	Leask *et al.* (1973)
Sodium	135–146 mmol/litre	Leask *et al.* (1973)
Potassium	3.6–5.2 mmol/litre	Leask *et al.* (1973)
Chloride	97–108 mmol/litre	Leask *et al.* (1973)
Bicarbonate	20–31 mmol/litre	Leask *et al.* (1973)
Urea	3.9–9.9 mmol/litre	Leask *et al.* (1973)
Creatinine	52–159 μmol/litre	Leask *et al.* (1973)
Calcium (men)	2.19–2.59 mmol/litre	Leask *et al.* (1973)
Calcium (women)	2.18–2.68 mmol/litre	Leask *et al.* 91973)
Phosphate (men)	0.65–1.23 mmol/litre	Hodkinson unpublished
Phosphate (women)	0.85–1.33 mmol/litre	Hodkinson unpublished
Alkaline phosphatase	3.6–14.5 kAU	Hodkinson and McPherson (1973)
Serum T4	58–128 nmol/litre	Hodkinson and Denham (1977)
T3-uptake (men)	93–122%	Hodkinson and Denham (1977)
T3-uptake (women)	95–125%	Hodkinson and Denham (1977)
FTI	0.58–1.24	Hodkinson and Denham (1977)
Random blood sugar	3.4–9.3 mmol/litre	Denham (1972)
SGOT	11–33 iu/litre	Hodkinson unpublished
LDH	58–145 iu/litre	Reed *et al.* (1972)
Bilirubin (men)	2–17 μmol/litre	Reed *et al.* (1972)
Bilirubin (women)	3–12 μmol/litre	Reed *et al.* (1972)
Uric acid	148–461 μmol/litre	Leask *et al.*, 1973

CHAPTER ONE

Policies for Investigation of the Elderly

In the past, it was all too often assumed that investigation of the elderly was a waste of effort because accurate diagnosis was superfluous as nothing therapeutically useful was likely to be achieved. Whilst it is not perhaps entirely unknown for such views to be expressed today, the achievements of geriatric medicine have largely eliminated such pessimism. It has been clearly demonstrated that many diseases can be treated just as easily and as effectively in old age as in younger age groups. Accurate diagnosis has been shown to be the cornerstone of effective medical care in old age as in other age groups.

The usefulness of biochemical investigation is indeed greater in old age where clinical diagnosis is so often fraught with special difficulties. Patients are often not able to give such a helpful history, physical examination is more difficult and the interpretation of physical signs sometimes less clear. Multiple disease frequently complicates the diagnostic process. Furthermore, atypical presentations of disease are especially common in old age, the mechanisms of production of pain and fever often being impaired. Non-specific presentations of illness are particularly common and of the greatest importance (Hodkinson, 1973). An old person with serious but treatable disease may merely go into an insidious decline whose features are no more striking than deterioration in energy and the ability to look after oneself adequately, increasing frailty and perhaps mild confusion or memory impairment. Such vague symptoms are all too readily and all too often laughed off as being merely due to 'old age' or 'senility'. Yet non-specific presentations are common in such conditions as uraemia, diabetes, osteomalacia, hypothyroidism or thyrotoxicosis, all of which may be readily recognised by appropriate biochemical investigation and may then be amenable to effective treatment.

On the other hand it is important that excessive investigational zeal such as might make the patient's illness more unpleasant than it need be should be avoided. As so much biochemical investigation can now be achieved by analysis of quite a small volume of blood from a single venepuncture, this consideration seldom precludes at least partial biochemical investigation of severely ill elderly patients. However, careful consideration is necessary whenever more taxing procedures are involved, for example fasting blood specimens, repeated venepunctures or tests involving the administration of drugs, and the decision must be based on the likelihood that the test will materially benefit the patient.

Discretionary tests or profile tests?

The traditional preference for discretionary tests which is now being undermined by the impact of multi-channel analysis making profiles more economical has already been referred to briefly. However, traditional clinical practice has never kept to an entirely discretionary use of tests, for example analysis of the urine for albumin and sugar is essentially a screening method in the sense of tests 'performed without a clinical indication' (Holland and Whitehead, 1974). Nevertheless, the general acceptance of profiles when they are offered to clinicians perhaps reflects the fact that they typically include the tests most often requested on a discretionary basis (Knox, 1974). They also confer advantages in that the clinician gets the results more quickly and needs to send a smaller amount of blood. Increasingly clinicians will have the decision in favour of a profile of tests made for them by the march of events in their hospital laboratory. The traditional neglect of the use of tests in a screening manner carries with it the risk that clinicians may fail to derive the full benefits of a profile. Clinical chemists have made many comments on the difficulties clinicians find in using profiles but have seen the solutions mainly in terms of improvements that the laboratory might make in presenting the results (Reece, 1974; Carter *et al.*, 1974; Whitehead and Wootton, 1974). Barnett and his colleagues comment (Barnett *et al.*, 1970), 'although laboratory problems in testing are still formidable, they are being solved far more rapidly than are the problems relating to the medical usefulness of the test results'.

I believe that the full potential usefulness of biochemical profiles will only be realised when clinicians learn new skills which cannot be achieved by reliance on spoon feeding – no matter how sophisticated or well intentioned – by their laboratory colleagues. Investigation of the elderly is perhaps a particularly appropriate situation in which such 'enlightened self help' might be practised. The considerable prevalence of treatable diseases which may be present atypically or non-specifically makes it essential that the clinician should master the screening as well as the discretionary methods of evaluating tests results. Furthermore, the frequent disturbances of test results by disease interactions, renal impairment or other metabolic disturbances or drugs mean that interpretation must be especially soundly based.

Sources of variation in laboratory test results

Interpretation of biochemical results from samples of blood or other biological fluids derived from patients would be a far simpler matter if only the effects of disease processes had to be considered. In practice, however, these changes have to be viewed against the often very considerable 'background noise' due to many other sources of variation. Some aspects of this variation may be avoidable, others may be allowed for or minimised whilst others again are intrinsic and unavoidable. Clearly, a test is far more powerful diagnostically if background variation is less. It is therefore very important that we should make every effort to reduce the variability of test results. Some avoidable extrinsic causes of alteration of test results will be considered in this chapter, whilst inevitable sources of variation will be dealt with in the next.

Avoidable errors

It is essential to avoid outright mistakes. Though such errors are often called 'laboratory errors' by clinicians, they mostly occur outside the laboratory! Mislabelling of a specimen is a common and important example. Confused elderly patients are particularly at risk as they may happily answer to an incorrect name. Erroneous mis-identification can only be kept to a minimum if some form of personal identification such as a wrist band is routinely used for inpatients and checking this must be a routine when venepuncture is performed. Having correctly identified the patient it is then essential that specimen containers are promptly labelled by the same person who undertook the venepuncture – an easier task now that patient labels are generally available in hospitals. Care must next be taken that the correct type of bottle (plain, heparinised, fluoride, etc.) is used or the analysis may be totally invalidated. Specimens must be taken to the laboratory without delay for inappropriate storage conditions or excessive length of storage may give rise to artefacts, for example due to haemolysis or escape of electrolytes from the red blood cells.

If excessive cuffing is used during venepuncture this will result in misleading elevation of macromolecular and cellular constituents of blood due to a process of ultrafiltration across the vessel wall. The use of venous cuffing should therefore be kept to a minimum though this may sometimes be something of a counsel of perfection when venepuncture is difficult in an ill, dehydrated elderly patient with fragile and mobile veins!

Anomalous results due to interference from drugs is another potential pitfall. The protean variety of ways in which iodine and its compounds could affect the determination of protein-bound-iodine, provided many examples when this test was widely used.

Errors in the laboratory itself are most often some kind of transcription error such as the transposition of results for different individuals or different tests or clerical errors involving the decimal point. Fortunately, such errors are less common now that computer systems are replacing manual systems in routine laboratory reporting.

Even when a correct report has been issued there remains the risk of misfiling in the wrong patient's records – no rare occurrence and all the more likely where patients with the same or similar surnames are involved.

When suspicious results do occur, the clinician should not dismiss them as 'laboratory error' and should instead consider the various possibilities discussed above. Where necessary, the assistance of laboratory colleagues should be enlisted whilst the trail is still warm.

Avoidable physiological variation

Some biochemical values are subject to circadian rhythms or other forms of diurnal variation. Figure 1 shows the particularly striking example of phosphate where both serum and urinary phosphate show a marked circadian rhythm with low morning and high evening

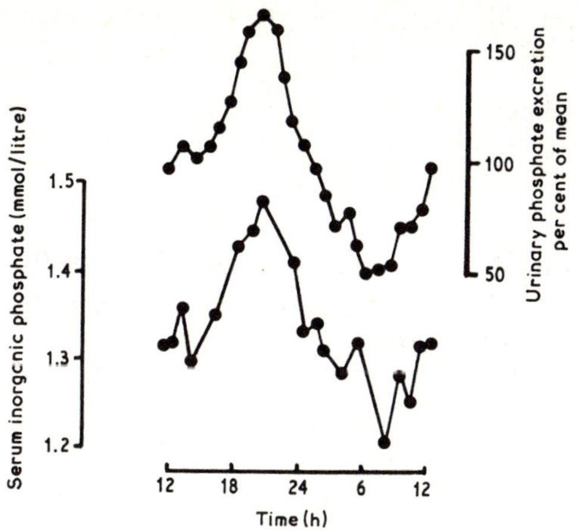

Fig. 1. Circadian variation of serum and urinary phosphate in healthy man (after Mills, 1966).

values. The value of biochemical investigations is thus greatly enhanced if routine specimens are habitually taken at the same time of day so as to effectively exclude this unwanted source of variation.

Similarly, posture may have considerable effect, particularly on cells, proteins or blood constituents which are protein bound, because of ultrafiltration effects similar to those associated with venous cuffing. A change from lying to standing may result in elevation of non-filterable blood constituents by as much as an eighth within 15 minutes whilst the reversed changes on lying down again take a little longer (Stoker, Wynn and Robertson, 1966). Again, the practical value of test results will be far greater if specimens are always taken in recumbency so as to avoid these variations.

Finally, biochemical results may be affected by meals, the effects on blood glucose being an obvious example. Here too standardisation of practice is called for, i.e. routine specimens should be taken at a standard time and with a constant relationship to the nature and time of the last meal.

Good practice in specimen collection

No amount of skilful interpretation can overcome the disadvantages stemming from poor specimen collection. The salient points of the foregoing sections can be summed up in a very few words – avoid mistakes by CARE in specimen collection, avoid unnecessary variation by STANDARDISATION.

Which tests are useful in the elderly?

The usefulness of individual tests will be more fully considered in the relevant later chapters. I would regard the following as the most useful of the commonly available blood tests:

serum sodium
serum potassium
blood urea
serum albumin
serum globulin
serum calcium
serum inorganic phosphate
serum alkaline phosphatase
serum T4 and T3-uptake (or ETR)
random (mid-morning) blood glucose
serum bilirubin
serum aspartate transaminase.

To these should be added the analysis of urine for protein, glucose and bile.

CHAPTER TWO

Principles of Interpretation of Tests

In order to make the best use of biochemical tests in diagnosing the presence of disease we need to have a basic understanding of their behaviour in healthy subjects. In particular we need to have an understanding of the intrinsic sources of variation of test results.

Causes of random variation

Analytical variation

There may be considerable differences in the analytical results for a chemical constituent where different chemical methods have been used or where the same method has been used by different laboratories. The differences are sufficiently great to have given rise to national systems of comparisons between laboratories aimed at reducing the discrepancies. These differences point to the need for considerable caution in accepting normal ranges or other findings from other laboratories, particularly when they relied on different analytical methods.

For the individual clinician, however, the analytical variations within one's own hospital laboratory are of paramount importance. Nowadays hospital laboratories make use of sophisticated methods of quality control and achieve far higher standards of reliability than were usual in the past. Nevertheless, though analytical variation can be minimised, it can never be wholly eliminated.

The nature of analytical variation will be considered in some detail for it affords our first encounter with the *normal distribution* (or *Gaussian distribution*), a distribution that holds a central place in statistics and to which we will need to make repeated reference.

We can most easily understand analytical variation if we consider what would happen if we carried out many repeat analyses for a particular constituent on aliquots of a single large sample. Even with the most careful technique one would not expect to get exactly the same result from each analysis. There must be a multiplicity of minor random factors not under one's control which can shift the result up or down. If we were to plot the results of a large number of analyses as a histogram we would find that it approximated to the symmetrical, bell-shaped curve of a normal distribution such as that shown in Fig. 2. An explanation of why the normal curve should describe this and other instances of random variation is beyond our present scope but is particularly well explained for the non-mathematical reader by Lewis (1966).

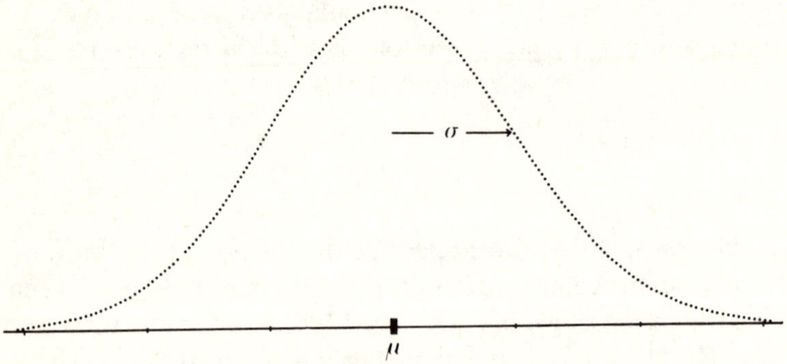

Fig. 2. The normal (Gaussian) curve showing the mean (μ) and standard deviation (σ). The scale is in standard deviation units.

The normal distribution

Each normal curve is characterised by simply its *mean* (i.e. the arithmetic average of all the observations) and its *standard deviation.* Standard deviation, mathematically, is the square root of the average of the squares of the deviations of observations from the mean. That is, if x represents observed values and N their number:

$$\text{mean } (\mu) = \frac{\Sigma x}{N}$$

$$\text{and standard deviation } (\sigma) = \sqrt{\left(\frac{\Sigma(\mu - x)^2}{N}\right)}$$

However, when, as is usually the case, mean and standard deviation are not derived from the universe of possible observations but estimated from a sample, these estimates are distinguished by using $\bar{x}$

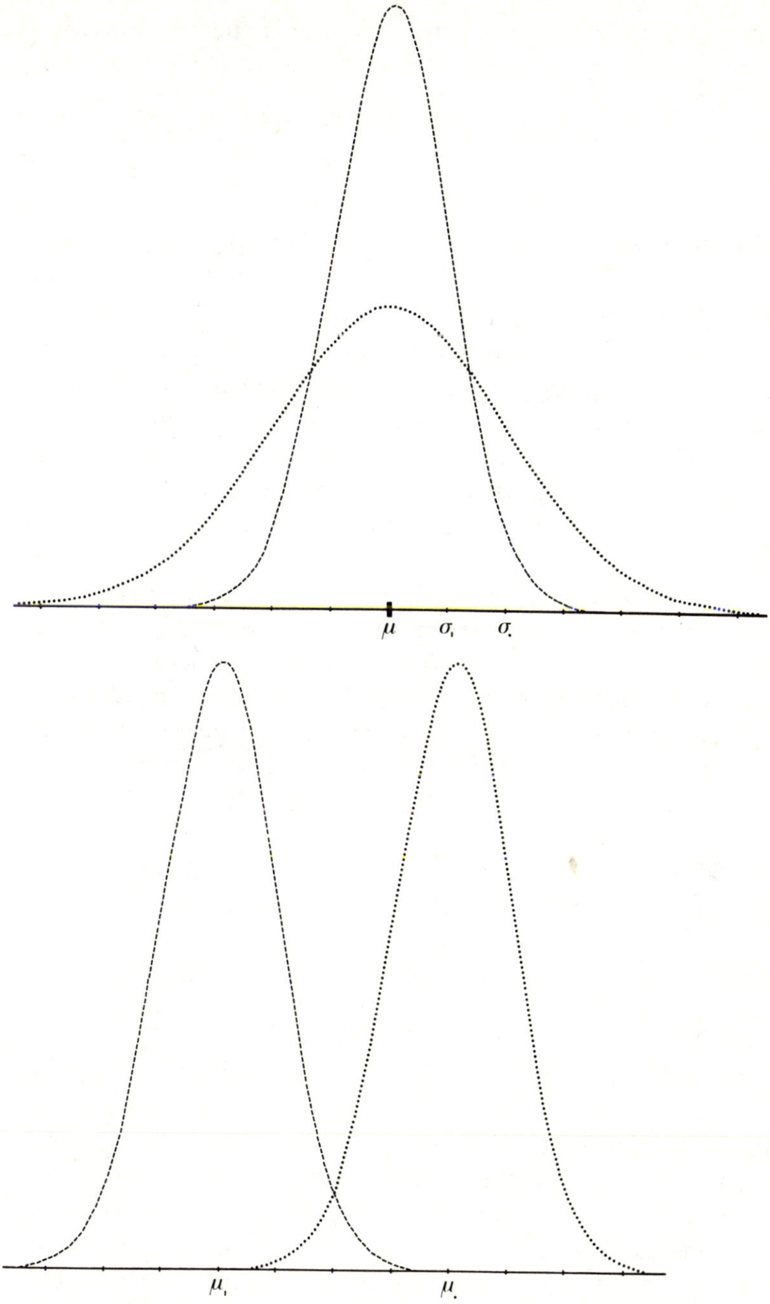

Fig. 3. Pairs of normal curves with (a) the same mean but different standard deviations and with (b) the same standard deviation but different means.

and s in place of μ and σ. Furthermore, as standard deviation is otherwise systematically overestimated in samples, $(N-1)$ replaces N in its calculation:

$$s = \sqrt{\left(\frac{\Sigma(\bar{x} - x)^2}{(N-1)}\right)}$$

The non-mathematical reader may find it easier to grasp these concepts visually. The mean gives the *location* of the centre of the normal curve whilst the standard deviation gives its *spread.* The standard deviation in geometrical terms is the distance from the mean to the *point of inflexion*, the point where the curve changes from convex to concave (Fig. 2). Two normal curves with the same mean but different standard deviations thus agree in regard to their location (Fig. 3a) whereas two curves with different means but the same standard deviation have the same spread but differing location (Fig. 3b).

The reader may wonder, at this point, why we should trouble to fit observations to a normal distribution. The answer to this question will, hopefully, become clearer when we have considered a number of examples but, in essence, the reason is that of greater convenience. We can describe a large collection of observations adequately by just two numbers, the mean and standard deviation, and having done so are then able to apply a great variety of useful statistical techniques which are based on normal distribution theory. When we cannot describe data by the normal distribution these methods are denied us and we are obliged to use less versatile and generally more cumbersome *non-parametric* methods.

Within batch and between batch variation

Common sense tells us that analytical variation is likely to be greater for analyses carried out in different batches than for those performed together in the same batch. If we carry out a large number of replicate analyses of the same specimen within one batch we can directly estimate within batch variation simply by calculating the mean and standard deviation of the results. If, however, our replicates are performed in many different batches, the greater variation then found when the standard deviation is calculated must be due to within batch and between batch variation combined. A fortunate property of the normal distribution allows us to quantitate the two contributions. This is that *variance* is additive, variance simply being the square of standard deviation. In other words we merely have to subtract within batch variance (i.e. the square of the standard deviation for within batch replicate analyses) from the combined variance (i.e. the square of the standard deviation found for replicates in many batches) to obtain between batch variance.

Within subject and between subject variation

Quite apart from any systematic variation such as that of a circadian rhythm, an individual's blood level of a given biochemical constituent also shows random variation. This within subject variation again is usually normally distributed. If we take multiple samples from a group of individuals so as to estimate a mean value for each, these means will vary and this between subject variation is also normally distributed as a rule. In practice, within and between subject variance would be estimated from replicate sample data using the technique of partitioning of variance we have already discussed under analytical variation.

The overall contributions of within and between batch variance and within and between subject variance to the variation of some common biochemical tests is shown in Table I which is based on data from Harris (1974) and Flynn and colleagues (1976). It will be seen

TABLE I

Relative contributions to variance in biochemical tests in healthy subjects

Test	Percentage of total variance			
	within-subject	between-subject	within-batch	between-batch
Serum albumin	21.3	57.7	4.8	16.2
Serum calcium	11.8	33.5	4.8	49.9
Blood glucose	30.2	50.7	0.9	10.5
Serum aspartate transaminase (females)	33.1	54.0	5.5	7.3
Serum inorganic phosphate	24.6	55.0	3.0	20.1
Serum potassium	47.0	32.6	3.6	16.8
Serum sodium	32.8	12.1	13.3	41.8
Blood urea	28.6	66.6	2.4	2.4

that, although there are differences in degree and a few exceptions, analytical variation is generally far smaller than individual variation. Between batch variance is usually larger than within batch and between subject variance greater than within subject.

Causes of systematic variation

We have already considered circadian rhythms as a cause of systematic variation. This at least can be avoided by standardisation. There are other unavoidable sources of systematic variation however, age and sex being the most important. Others include such factors as race, blood group, genetic factors and environmental and dietary effects.

Effects of age

Systematic changes with age have been shown to occur with many biochemical measurements (Roberts, 1967; Keating *et al.*, 1969). Clinically important age differences will be discussed more fully when individual tests are considered in later chapters and for the present we may look at a single example. Table II shows how serum albumin falls progressively with age (data for men from Keating *et al.*, 1969). The change affects only the mean, standard deviation remaining virtually constant throughout the age groups.

TABLE II

Changes of serum albumin with age in healthy men

Age (years)	Serum albumin (g/litre)	
	Mean	Standard deviation
20–29	40.03	2.94
30–39	39.25	2.95
40–49	38.47	2.95
50–59	37.68	2.94
60–69	36.90	2.94
70+	36.12	2.94

Effects of sex

Sex differences may also occur for a number of biochemical values (Roberts, 1967; Keating *et al.*, 1969). An example is serum phosphate which is considerably lower in old men as compared with old women. Figure 4 shows the idealised distributions for phosphate values in the two sexes in old age (Hodkinson, 1973a) – two normal distributions differing in means but not in standard deviation.

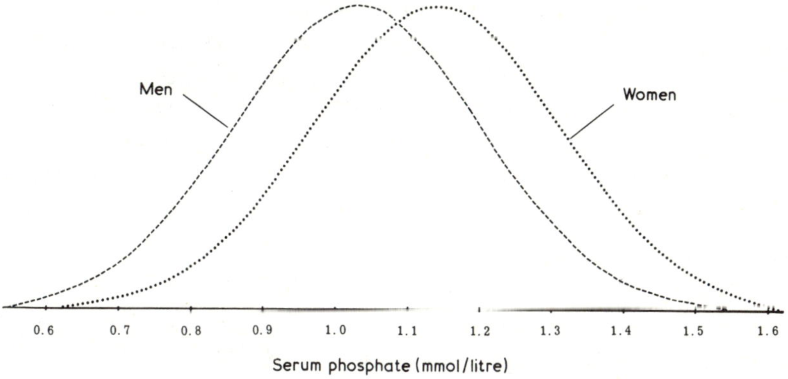

Fig. 4. Differing distribution of phosphate results in elderly men and women.

The log-normal distribution

So far we have talked only of the normal distribution as a description of the variability of biochemical results. However, a number of commonly measured biochemical values, for example urea, creatinine, glucose, bilirubin, alkaline phosphatase and aspartate transaminase, have distributions which are *log-normal* instead of normal. That is to say that the distributions are skewed to the right but become normal if the observations are replaced by their logarithms.

This *logarithmic transformation* of data is often viewed with considerable suspicion by clinicians who suspect that it in some way represents cooking the results! They may be reassured by the thought that pH results, for example, are already expressed in logarithmic form yet they would not feel uneasy if they found them to be normally distributed. Lewis (1966) gives a readily understood explanation as to why some types of measurement should assume log-normal distributions. Here, it is perhaps enough to stress the entirely practical aspects of logarithmic transformation; that it allows us to use normal statistics where otherwise we would be unable to do so.

'Normal ranges'

We have seen how the main contributions to random variation of biochemical measurements, analytical, within- and between-subject variation, are normally distributed. So the results from a homogeneous sample of individuals are also normally distributed. Homogeneity is stressed as, in its absence, the underlying normality of the distribution may be distorted by systematic variation of one kind or another. Figure 5 shows how non-normality may result from mixing two similar normal populations. Such effects may readily occur where there are systematic sex or age differences. Failure to appreciate the causes of distortions of this kind has tended to bring the use of normal statistics in biochemical work into undeserved disrepute (Elveback, Guiller and Keating, 1970) and to support the suggestion that they should be generally replaced by less versatile non-parametric methods. However, where care has been taken to ensure homogeneity with regard to age, sex and health status, the close fit of observed biochemical values to normal distributions (or log-normal ones) has been clearly demonstrated (Flynn *et al.*, 1974).

The proportion of sample results drawn from a normal distribution which can be expected to lie within a given deviation from the mean can be predicted, being given by the area under that portion of the normal curve.

Thus, 95% of observations may be expected to lie within ±1.96 standard deviations from the mean. These limits ($\bar{x} \pm 1.96\,s$) are

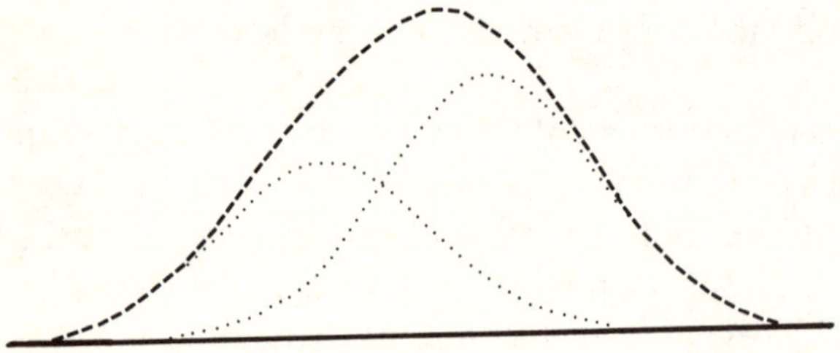

Fig. 5. A non-normal distribution resulting from the mixing of two normal populations.

those of the conventional 'normal range', the mean and standard deviation having been found from an appropriate sample of subjects. Normal range is a somewhat unfortunate term because of confusions engendered by the different possible meanings of normal. Normal in the sense of Gaussian is implied but this is readily coloured by the alternative meanings of 'non-pathological' and 'usual'. All too easily one may come to think of values outside the normal range as 'abnormal' despite the fact that 5% of the population from whom the normal range derives must by definition be expected to have such values. All too often normal ranges are used in this way to classify results as normal/abnormal. When dealing with multiple results such use (or abuse) of normal range gives rise to particularly serious problems. Table III shows the expected percentages of normal subjects who would have given numbers of 'abnormal' results with increasing numbers of tests. When the number of tests passes thirteen, less than half of the normal subjects will have entirely 'normal' tests on average.

It would be better if the confusing term normal range were replaced, perhaps by '95% range in health' which, though more cumbersome, would be explicit and would remind us that 5% of healthy subjects will have results outside the range. The term *reference range* also has some advantages as it underlines the need to determine 95% range from an appropriate *reference population*, that

TABLE III

The probabilities of given numbers of 'abnormal' test results for multiple independent tests performed in the healthy subject.

Total number of tests	Expected percentages				
	No abnormals	1 abnormal	2 abnormal	3 abnormal	4 abnormal
5	77.4	20.4	2.1	0.1	0.0
10	59.9	31.5	7.5	1.0	0.1
15	46.3	36.6	13.5	3.1	0.5
20	35.2	37.7	18.9	6.0	1.3
25	27.7	36.5	23.1	9.3	2.7
30	21.5	34.0	25.9	12.7	4.5
40	12.9	27.1	27.8	18.5	9.0
50	7.7	20.2	26.1	22.0	13.6
100	0.6	3.1	8.1	14.0	17.6

is from healthy subjects of appropriate age, sex and race and using the same laboratory methods.

The choice of 95% limits for the conventional normal range is, of course, entirely arbitrary. We have seen how difficulties increase when multiple test results are classified as normal/abnormal by means of normal ranges. The use of wider ranges has been suggested as a way of dealing with this problem. Figure 6 shows how the probability of having a result outside given limits falls away quite steeply beyond 2 standard deviations. However, no matter how much wider limits are taken, there will always be a residual probability of the healthy person having a result outside them whilst the wider

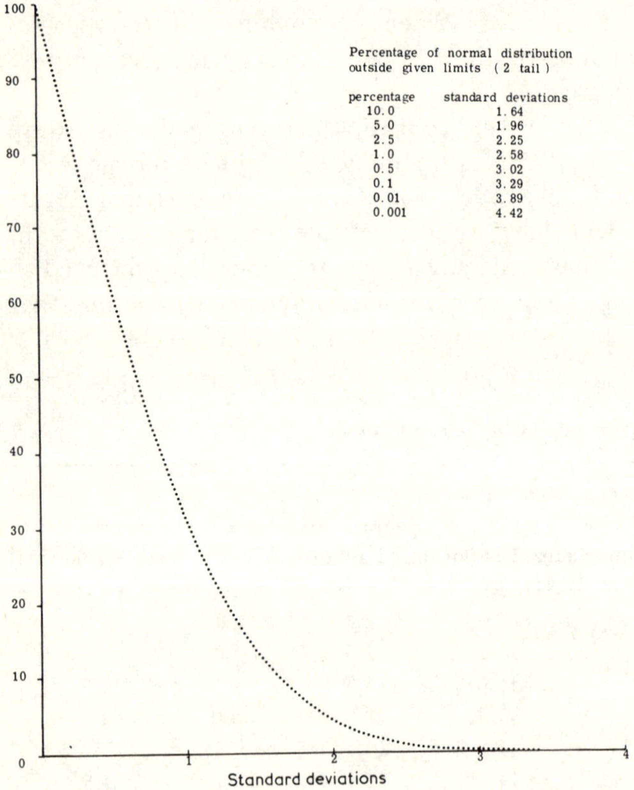

Fig. 6. The percentage of the total area under the normal curve which is outside ± x standard deviations from the mean.

limits will incur an increasing risk of accepting abnormal results as normal. Tinkering with the normal range cannot overcome its inherent disadvantages. Perhaps normal range should be used only as a familiar and reasonably convenient way of memorising the distribution of test results (although it is no easier than remembering mean and standard deviation) and we should look for more effective ways of interpretation.

Range in health and range in disease

So far we have only considered the distribution of test results in health. We shall see that a knowledge of the distribution of results in disease is equally necessary if we are to interpret test results satisfactorily. Unhappily, though much is written of normal ranges in health, the distribution of test results in disease is far more rarely made available.

Typically, where a test is altered by a disease, the distribution of test results in disease has a similar standard deviation but a different mean. Figure 7 shows a hypothetical example, the difference in means of 3 standard deviations being somewhat on the generous side; we shall see that single tests often have greater overlap when we consider actual examples in later chapters.

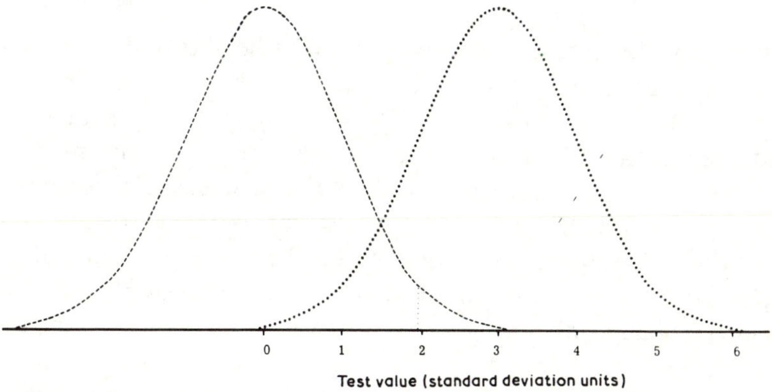

Fig. 7. Overlapping distributions for hypothetical test results in health (left) and disease (right).

Test interpretation in a discretionary situation

Let us suppose that the hypothetical test in Fig. 7 is elevated in a given disease which, on clinical grounds, the clinician strongly suspects and therefore orders the test. Clinicians do not usually express their diagnostic convictions in numerical form but let us further suppose that in this patient's case he believes there to be an equal chance of the patient having the disease or not. In other words, he estimates the *prior probability* for the disease at 0.5, prior probability being the probability before the test result is known.

The clinician will of course know the normal range for the test but will usually have no detailed knowledge of its range in the disease – he will merely have his general recollection of results in previous patients in whom this diagnosis was made. He is likely to rely almost exclusively on his knowledge of the normal range and, remembering that only $2\frac{1}{2}$% of results in the healthy lie above the upper normal limit, he is probably going to regard any result above this limit as making the diagnosis practically certain. Conversely, although he may be somewhat dubious if a result is just within the upper limit, he is likely to regard a result within the normal range as making the diagnosis very unlikely. How well is the clinician served by this rough and ready application of normal range in this particular case?

We can best answer this question by looking again at Fig. 7. This accurately represents the diagnostic situation – we believe the possibilities of having the disease and of not having the disease to be equal so we represent them by the two equal sized distribution curves. At any given test value, therefore, the heights under the two curves at that point will show us the respective chances of having the disease or not. In fact, these *relative probability densities*, as they are termed, can easily be calculated from the formula of the normal curve as is shown in Appendix I. For any test value a *posterior probability* for the disease can therefore be obtained readily by dividing the disease probability density by the combined probability densities for disease and non-disease. Figure 8 shows how the posterior probabilities change for test values between 0 and 4 standard deviations.

Posterior probability follows a sigmoid curve with its inflexion (corresponding to 0.5 probability) at 1.5 standard deviations, the value at which the two curves in Fig. 7 intersect. This value, called

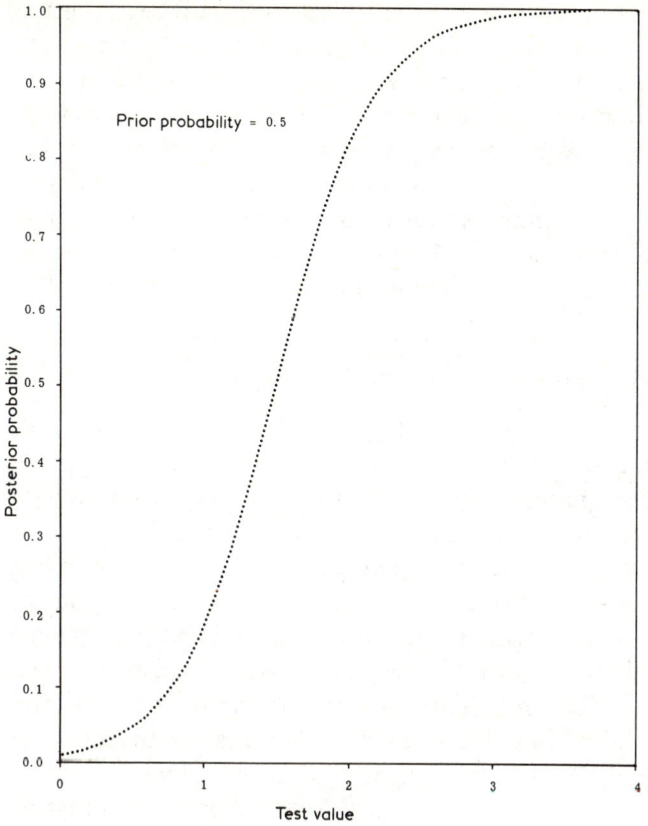

Fig. 8. Posterior probability for disease from values of the hypothetical test when prior probability is 0.5.

the *critical limit* (Murphy and Abbey, 1967), provides a more satisfactory guide to interpretation than does the upper normal limit. Values below the critical limit are associated with major probabilities against disease and values above the limit with major probabilities for disease. Unfortunately critical limit varies according to prior probability and would need to be estimated in each particular case, a fact which considerably limits its practical utility.

If we take a 'practically certain' diagnosis to be one with a posterior probability of 0.95 or more, Fig. 8 shows that this level of certainty of diagnosis of non-disease or disease obtains only below 0.5 and above 2.5 standard deviations. The rather wide zone of lesser degrees of certainty would include 30.8% of possible test results. A result at exactly the upper normal limit (1.96 standard deviations) is associated with a posterior probability of only 0.8 for disease. The clinician's decisions based on normal range alone would thus be perfectly sound in the 34.1% of results below 0.5 standard deviations and in the further 34.1% above 2.5 standard deviations but less satisfactory for intermediate values. Here his views are too definite. He underestimates the inherent uncertainties and fails to recognise that values between 1.5 and 1.96 standard deviations, though within the normal range, imply a probability for disease of greater than 0.5.

Non-discretionary use of tests

Normal range is thus an imperfect guide to interpretation in the discretionary situation though not disastrously bad. Its inadequacies become far more glaring when we turn to a non-discretionary situation. If we return to the same hypothetical test and disease we now have to consider a situation where prior probability is low. We are now using the test to screen patients in whom we find nothing clinically to suggest the disease, previous experience having shown a certain prevalence of clinically unrecognised disease in such a patient population. Our prior probability is our current estimate of this prevalence rate and will usually be of the order of a few percent. For our example we shall adopt the slightly generous value of 0.05 for prior probability. Figure 9 illustrates the new diagnostic situation with the far larger non-disease population and small disease popu-

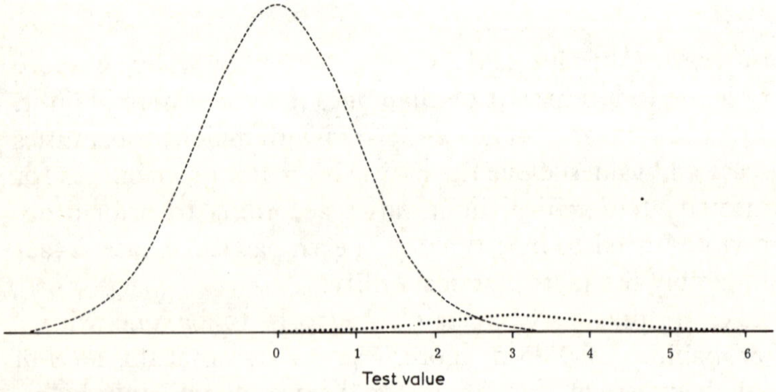

Fig. 9. Overlapping distributions for the hypothetical test when the prevalence of disease is only 0.05.

lation with heights in the ratio 0.95 : 0.05. The critical limit, corresponding to the point of intersection of the two curves has moved a standard deviation to the right from 1.5 to 2.5 standard deviations. Figure 10 shows the altered curve for posterior probabilities – a very similar curve to that of figure 8 but displaced one standard deviation to the right. The upper normal limit of 1.96 standard deviations now corresponds to a posterior probability of disease of only 0.175 and is clearly a totally unsuitable guide to interpretation in this screening situation. Critical limit is now seen to be far better than the normal limit. Its use would lead to 99.4% of those without disease being correctly classified and 69.1% of those with disease. However, 0.95 probability or better would be achieved

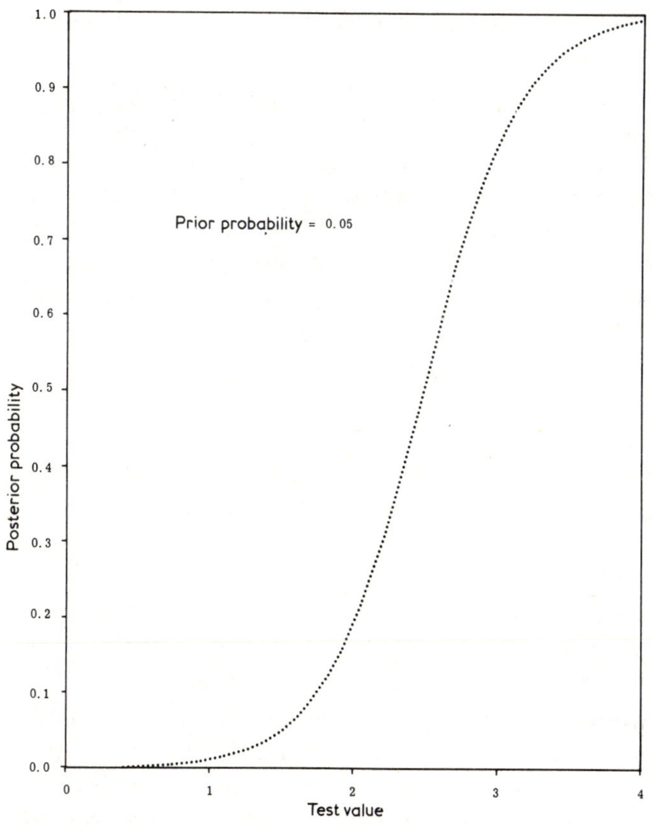

Fig. 10. Posterior probabilities of disease from values of the hypothetical test when prior probability is 0.05.

for 93.3% of those without disease but only 30.8% of those with disease. The best policy would thus seem to be to utilise the actual estimate of posterior probability, proceeding to further investigation where this falls short of the required level of certainty.

Repeat tests and alternative tests

When a test gives an uncertain result we are often tempted to repeat it. How much further information are we likely to gain by doing this and how should the new result be interpreted. Unfortunately, we cannot adopt the posterior probability derived from the first test as our new prior probability and use the repeat test to calculate a further posterior probability because the two results are not *independent.* This is because, being two results for the same test in the one person, the contribution of between-subject differences to variation is eliminated.

When we are using the results of two tests on our patient we must therefore compare their mean with the means of duplicate samples for our disease and non-disease populations. Fortunately, we are able to do this quite readily. The variance of the mean of a number of samples from a normal distribution is given by the variance of single samples divided by the number. In the present case therefore within-subject variance will be halved whilst analytical and between-subject variance remain unchanged. If within-subject variance is 0.3 for our test (a reasonably representative value as can be seen from the actual examples given in Table I), variance will be reduced to 0.85 and standard deviation to its square root, 0.92. We can now proceed in exactly the same way as we did for single samples but using the mean of the two samples and employing 0.92 instead of 1.0 as standard deviation for the two populations.

To get a clearer idea of the practical usefulness of repeating a test, let us consider what would happen if the 30.8% of tests which failed to achieve 0.95 probability in our first example (where prior probability was 0.5) were repeated. It can be estimated that the duplicate tests would reach a diagnostic probability of 0.95 in 43% of cases. In other words, whilst single tests reached this level of decisiveness in 69% of cases, duplicate tests raise this to 82% – a rather modest improvement in what may be regarded as a generally favourable case. Further repeats are of very little additional value for whilst the first repeat halved within-subject variance, the next removes only a further sixth ($\frac{1}{2} - \frac{1}{3}$) and the next again a twelfth ($\frac{1}{3} - \frac{1}{4}$).

More useful gains in diagnosis must therefore rely on the use of alternative tests. The ideal would be a test whose results were totally independent of those of the first test. We could then make the posterior probability from the first test the prior probability for the second and calculate a new posterior probability in the usual way.

Unfortunately, totally independent tests are rare in practice. Though many are independent in health, most are *correlated* in disease, that is to say that they tend to vary either in proportion or in inverse proportion to each other. We then need to combine the test results in some way, just as we combined repeat tests by taking their average.

Sometimes there may be an obvious way in which to combine tests. For example, there is a strong theoretical justification for combining T4 and T3-uptake results as the free-thyroxine-index as this more accurately reflects the levels of free thyroxine and hence thyroid status. This single result can then be used to calculate a posterior probability in the usual way (Chapter 7). Otherwise, we need to find some other sound basis for the combination of tests. This is best provided by *discriminant function analysis.* The reader is referred to statistical texts for details of this complicated technique which can, however, be readily carried out when a computer is available and standard program packages (Dixon, 1967) are available. In terms of our present requirements, given the individual test values for two groups of subjects with and without the disease, discriminant function analysis finds the linear combination of weighted test values, called the *discriminant function*, which gives the widest possible separation between the means of the two groups. For a series of test results, $x_1, x_2, \ldots, x_n$, the discriminant function is of the form, $f_1x_1 + f_2x_2 + \ldots + f_nx_n$; f_1, f_2, etc., being the individual weighting factors found by the analysis. As the discriminant function combines multiple values which are normally distributed, the function will also be normally distributed. Its standard deviation is always standardised at unity by the analysis. The discriminant function thus provides exactly what we need for the calculation of posterior probability from a set of results. Though hitherto applied rather infrequently to problems of medical diagnosis, discriminant functions offer great possibilities. Their application to some specific diagnostic problems will be illustrated in a later chapter (Chapter 6).

The need to calculate the discriminant function or derive it from a complicated nomogram (e.g. Fraser *et al.*, 1971) has hitherto been a deterrent. However, now that relatively inexpensive programmable pocket calculators are available, discriminant functions can quite easily be put to use in the everyday clinical situation.

CHAPTER THREE

Misleading Results due to Illness or Medication

Misleading test results due to the general effects of illness or to technical interference or biological side-effects of medication may occur at any age. Bradwell and his colleagues (1974) found unrecognised effects of this kind accounted for a substantial proportion of all misleading abnormalities of hospital profile results. The opportunities are especially great in the elderly. Illness is often severe and associated with multiple pathologies whilst treatment all too often involves a multiplicity of drugs. In consequence, there are many opportunities for test disturbances which hinder rather than help diagnosis. The clinician investigating elderly patients needs to be constantly vigilant so as not to be misled. The more important pitfalls will be mentioned in the context of individual tests in later chapters. Here, a more general account of the main phenomena will be given.

Effects of impaired renal function

Impairment of renal function due to pre-renal, renal or post-renal causes is very common in the elderly patient. For example, the 95% range for blood urea in admissions to the geriatric department of Northwick Park Hospital is 3.1–20.6 mmol/litre (19–124 mg/100 ml) (Hodkinson, 1973c) as compared with a range of 3.6–10.0 mmol/litre (22–60 mg/100 ml) for the well elderly (Leask *et al.*, 1973). In the same way, creatinine is often raised in the elderly patient (Bahemuka *et al.*, 1974). Elevation of urea and creatinine give indispensable information regarding renal function, both showing a close correlation with the glomerular filtration rate of elderly patients (Denham *et al.*, 1975). However, other tests which add nothing to the assessment of renal function are also affected when it is impaired. Thus, phosphate and uric acid tend to be elevated and may be of very little practical value when renal function is disturbed. Figure 11 shows how the distribution of phosphate results is severely skewed to the right in patients admitted to the Northwick Park Hospital geriatric department. Urea, phosphate and creatinine are significantly inter-correlated in these patients though they are not correlated in health (Flynn *et al.*, 1976).

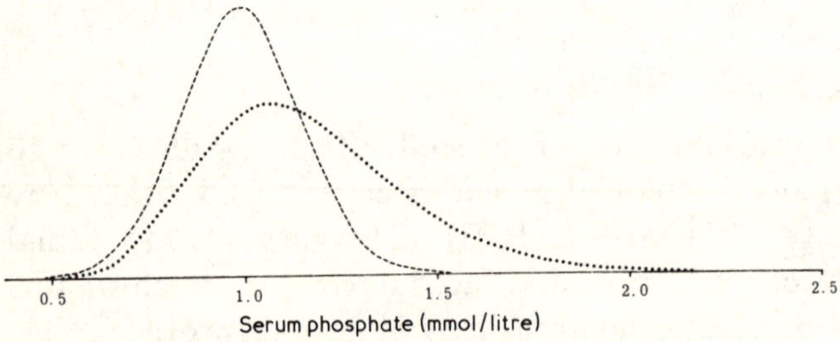

Fig. 11. Effects of renal impairment on serum phosphate: distribution of results in healthy elderly men compared with the higher skewed distribution in elderly male patients, many of whom have impaired renal function.

Dehydration

Dehydration is a common feature of severe illness in old age. Apart from contributing to test abnormalities because of consequent pre-renal uraemia, dehydration also elevates the levels of many blood constituents because of haemoconcentration. The plasma proteins may be elevated or an abnormally low level may be masked. Excessive cuffing during venepucture may exaggerate the effects of dehydration. Dehydration and cuffing effects appear to be quite common in hospital practice (Hodkinson, 1973c).

Protein binding phenomena

Many important blood constituents are substantially bound to plasma proteins. Calcium is mainly bound to albumin but also to a lesser extent to globulins. Albumin also binds magnesium, cortisol, free fatty acids and bilirubin. Thyroid hormones are chiefly bound to a specific carrier protein, thyroxine-binding-globulin. Other specific carrier proteins bind iron and the fat soluble vitamins whilst cholesterol is bound to β-globulin.

Severe illness may be accompanied by a fall in the levels of albumin and the other carrier proteins. This appears to be a non-specific effect related to the severity of illness rather than its type and is presumably due to impaired protein synthesis as a result of the constitutional upset. Lowered protein levels are particularly common in elderly patients where they have a considerable adverse prognostic significance (Jefferys *et al.*, 1972; Hodkinson, 1975).

Low carrier protein levels can have a considerable impact on the measured levels of bound substances. Results in the elderly patient may often be grossly misleading unless some means of correction for protein alterations is utilised. Calcium needs to be corrected for variations in albumin and globulin as shown in Chapter 6, iron needs to be interpreted in terms of the percentage iron saturation and thyroid test results as a free-thyroxine-index (Jefferys *et al.*, 1972).

Disturbances may also occur because of elevation of carrier protein levels due to dehydration, excessive cuffing or disease such as myeloma. Drugs may also give rise to either elevation or lowering of protein levels or may compete for binding sites: examples are given below.

Failure of homoeostasis

Many homoeostatic mechanisms become progressively less efficient with increasing age and so major disturbances more readily occur in illness in old age. For example, Mukherjee and his colleagues (1973) have shown that apparently well old people may have defective osmoregulation. Severe disturbances of serum osmolality are therefore common in ill old people (Denham, 1973).

Test disturbances due to treatment

Therapy may affect biochemical test results in a bewilderingly wide variety of ways. Occasionally the disturbance may arise from a therapeutic method – elevation of serum creatinine phosphokinase or transaminases after intramuscular injections for example – but is most usually due to the use of drugs.

The present account can be no more than a bald outline of some of the main mechanisms by which drugs may give rise to abnormal test results. Young and his colleagues (1972) list more than 9000 examples of such effects. Their valuable paper may be consulted for more detailed information and for references to the literature.

We may divide drug effects into two broad categories, *in vitro* interference with the laboratory test and *in vivo* phenomena affecting the patient.

Interference with laboratory tests

The following examples illustrate just a few of the many ways in which drugs may interfere with biochemical analysis. The clinician can only hope to be familiar with the more important of the many potential pitfalls. Beyond this, he must rely on a high index of suspicion that drug effects might be responsible whenever he is confronted by an unexpected and inexplicable test result. In such cases he should make the fullest use of the help which his laboratory colleagues may be able to give.

Unwitting contamination is perhaps the most obvious way in which drugs may invalidate test results. Protein-bound-iodine estimations provide a wealth of examples. Commonly, up to 5% of results are elevated by exogenous iodine from cough mixtures, asthma remedies, enterovioform, radiological contrast media and a whole host of other medicaments which contain iodine.

Enzyme inhibition by drugs may decrease the enzyme activity measured in the laboratory. Furadantin or ethionamide may lower alkaline phosphatase or lactic acid dehydrogenase in this way.

Altered dye binding due to drugs may invalidate analyses which use colorimetry of dyes bound to proteins in their estimation. Thus, albumin estimated by the H.A.B.A. dye binding method may be spuriously high because of heparin therapy or too low with salicylate treatment.

Colorimetric analyses may give results which are too high when drugs cross react with the reagents. So, where the Jaffé reaction is used to measure creatinine for example, L-Dopa and ascorbic acid are chromogenic and give rise to spuriously high results.

Fluorescence techniques may similarly be insufficiently specific. Fenfluramine, spironolactone and fusidic acid may lead to abnormally high cortisol results when the sulphuric acid fluorescence technique is used.

Oxidation–reduction reactions on which other analyses are based may be invalidated by reducing substances such as ascorbic acid, fructose or glucose which have been used in treatment. The older analyses for glucose which depended on reduction of alkaline copper solutions could be falsely elevated in this way because of lack of specificity.

In vivo effects of drugs

Some drug effects are readily appreciated as they merely represent the effects of *over-treatment.* Common examples are hypokalaemia and hyponatraemia due to diuretics and hypoglycaemia from oral diabetic agents or insulin.

Precipitation of latent disease also affords a readily understood mechanism for the alteration of test results. Thus glycosuria and hyperglycaemia due to unmasking of diabetes by thiazides or corticosteroids is quite commonly seen in elderly patients. These same drugs may also trigger off gout and give hyperuricaemia. Iodine compounds, e.g. radiological contrast media or potassium iodide given to block the thyroid during radio-iodinated-fibrinogen studies for thromboembolism, may lead to abnormal thyroid function tests because of transient thyroid suppression or, not infrequently in the elderly, because thyrotoxicosis is precipitated (Denham and Himsworth, 1974).

Hepatotoxicity may be a serious adverse side-effect of drugs and thus lead to abnormal liver function tests. The cholestatic effects of chlorpromazine are perhaps the most familiar example. However, many other drugs may sometimes give less obvious and more transient disturbances, perhaps giving merely modest enzyme elevations, which are less readily recognised. The anti-tuberculous drugs, cephalosporin antibiotics, monoamine oxidase inhibitors, methyldopa and many other drugs may behave in this way.

Diminished hepatic excretion of bilirubin may occur without liver damage because of competition for the excretory mechanisms from radiological contrast media. Bilirubin will thus be elevated but other liver function tests will be unaffected.

Nephrotoxicity may also occur as a serious drug side-effect. Kanamycin and tetracyclines are important examples of drugs which may elevate urea and creatinine in this way.

Enhanced renal excretion may result from drug actions on the other hand. For example, uric acid levels may be reduced because of the uricosuric effects of drugs such as aspirin, phenylbutazone and allopurinol. Diuretics may lower the levels of potassium and sodium and also sometimes of chloride, bicarbonate or calcium.

Enzyme induction may result when drugs have been given for some time. This increased activity of the liver microsomal enzyme systems results in the more rapid breakdown of metabolically active substances. So, for example, in patients on prolonged anticonvulsant therapy, vitamin D is more rapidly converted to inactive metabolites and osteomalacia with raised alkaline phosphatase and lowered calcium and phosphate may result.

Endocrine and hormonal mechanisms may account for a variety of other test abnormalities due to drugs. The mineralocorticoid side effects of corticosteroids or carbenoxolone may result in hypokalaemia. Conversely, spironolactone may give potassium elevation. Some drugs, chlorpropamide being a striking example, may give rise to the syndrome of inappropriate ADH secretion characterised by severe hyponatraemia.

Other drugs may act on the pituitary of hypothalamus and give rise to a variety of metabolic effects. A common example in elderly patients is the elevation of serum T4 which may follow L-Dopa administration.

Oestrogens antagonise the actions of parathormone and give a considerable reduction of serum phosphate and a less pronounced fall in serum calcium.

Protein binding may be affected either by competitive binding of drugs or by alterations which they induce in the levels of the carrier proteins. Competitive binding leads to lowered levels of the natural bound substances. Thus, for example, phenytoin or phenylbutazone may displace thyroid hormones from thyroxine-binding-globulin or aspirin, penicillin or sulphonamides compete with bilirubin and cortisol for albumin binding sites.

Carrier protein levels are increased by oestrogens, clofibrate and perphenazine and lowered by androgens and anabolic steroids. These drugs may thus have striking effects on thyroid function tests in particular, thyroxine-binding-globulin being particularly affected.

Interpretational difficulties due to multiple pathology

Multiple pathology is a typical feature of illness in old age. How may this affect test interpretation? This may be best shown by a brief consideration of the use of alkaline phosphatase as a screening test for the recognition of osteomalacia. This has considerable value in younger age groups but is of very little help in the elderly patient. This is because of the very high prevalence of other diseases giving rise to elevation of alkaline phosphatase. In a series of over five hundred patients admitted to a geriatric department, osteomalacia proved to be a relatively infrequent cause of a raised alkaline phosphatase (Hodkinson and McPherson, 1973). It accounted for only 14% of raised values in women and for none in men. Liver disease, fractures, Paget's disease and rheumatoid arthritis were all more frequently the causes of elevation of the enzyme. Furthermore, a considerable number of patients had more than one cause for alkaline phosphatase elevation. So, whilst a normal alkaline phosphatase is of considerable value in excluding the diagnosis of osteomalacia, a raised alkaline phosphatase result is of very little diagnostic value in the elderly patient.

CHAPTER FOUR

The Plasma Proteins

The plasma proteins are a complex mixture of many different proteins with widely varying functions. The three main plasma proteins – fibrinogen, albumin and the globulins – will be considered in this chapter. The plasma enzymes are considered separately in Chapters 6 and 9.

Fibrinogen

Fibrinogen accounts for approximately $2\frac{1}{2}-5\%$ of the total plasma protein concentration. It has a very high molecular weight (approximately 340 000). Its function is in clot formation when it is converted to insoluble fibrin. It is therefore virtually absent from serum so that the term 'serum proteins' refers to the remaining plasma proteins, principally albumin and the globulins. The latter are often estimated in clinical practice whereas fibrinogen is not. However, fibrinogen is a major determinant of the erythrocyte sedimentation rate (ESR), a widely used test which is discussed below.

Albumin

Albumin is the most important plasma protein quantitatively, accounting for more than half of the total concentration in health. It has a relatively low molecular weight of approximately 65 000 and this coupled with its high concentration make albumin the main contributor to the colloid osmotic pressure of the plasma whilst the globulins with generally much higher molecular weights contribute only 15% of the total on average.

Apart from its role in maintaining colloid osmotic pressure, albumin has several other functions. Important amongst these are its transport functions. Bilirubin, uric acid, free fatty acids, calcium and other divalent ions, cortisol and the thyroid hormones are some of the many endogenous substances bound by albumin. Many drugs are also bound by albumin. The protein also has important buffering properties associated with its ability to loosely bind monovalent ions, making a substantial contribution to the total buffering capacity of the plasma.

Some 7–11% of the total serum albumin is destroyed and replaced daily. Synthesis appears to be confined to the liver whilst catabolism appears to take place mainly in the gut, kidneys and liver.

The globulins

The globulins are a highly heterogeneous group of proteins with widely varying chemical structures and functions. They are grouped together largely for historical reasons, the globulins being all those proteins which were thrown out of solution by 'salting out' and other precipitation methods first used to fractionate the serum proteins. These methods are no longer generally used in clinical chemistry but the old usage continues – total serum protein is measured by dye-binding or biuret methods and albumin by dye-binding methods and the difference between the two values is reported as serum globulin.

A large number of individual globulins have been identified but, with the exception of the measurement of total iron binding capacity which estimates transferrin, individual protein concentrations are seldom determined in clinical practice. Electrophoresis of the serum proteins is quite commonly used to distinguish different patterns of change in the globulins, however. The paper or cellulose acetate strip can be stained and then scanned by densitometer to quantitate peaks representing albumin and four globulin groups, the α_1, α_2, β and γ-globulins (Fig. 12). In addition, a small tail to the cathode side of the large albumin peak due to pre-albumin may be recognisable.

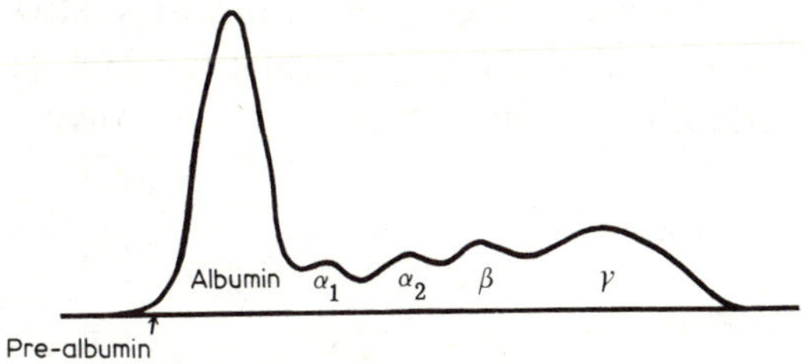

Fig. 12. Serum electrophoresis: the albumin and globulin peaks as seen in health.

Pre-albumin, though its name might suggest otherwise, is a globulin with a glycoprotein structure and a low molecular weight of approximately 61 000. It is a transport protein for the thyroid hormones but is less important than thyroxine-binding-globulin (TBG). Unlike the latter protein, it binds tri-iodothyronine (T3) more powerfully than thyroxine (T4).

The α_1*-globulins* are a heterogeneous group which includes the α_1-lipoproteins, α_1-glycoprotein, TBG and α_1-antitrypsin. The latter is quantitatively the most important of the group and acts as an inhibitor of trypsin and other proteases. The function of α_1-glycoprotein is uncertain and it may be a tissue breakdown product. The α_1-lipoproteins transport cholesterol, glycerides, phospholipids and vitamin A.

The α_2*-globulins* are also a mixed group. The function of α_2-glycoprotein is uncertain. Ceruloplasmin strongly binds copper and functions as an oxidase whilst the haptoglobulins bind haemoglobin and α_2-macroglobin acts as a protease inhibitor and possibly as a transport protein for insulin.

The β*-globulins* include the pre-β and β-lipoproteins which carry triglycerides, phospholipids and cholesterol, but also fat soluble vitamins and hormones. Transferrin, the iron transport protein, and components of the complement system are also β-globulins.

The γ*-globulins*, unlike the α and β groups, are functionally homogeneous. They comprise the immunoglobulins produced by lymphoid tissues. The three main types are IgG, IgA and IgM immunoglobulins which may be distinguished by immunoelectrophoresis.

The main function of the immunoglobulins is as antibodies providing defence against infection. Antibacterial antibodies occur in IgG, IgA and IgM subgroups though principally in the IgG group where most antiviral antibodies are also found. IgA also contains antitoxins, isoagglutinins and antinuclear factors. IgM includes ABO blood group isoantibodies, rheumatoid factor, antinuclear factors and other autoimmune antibodies such as those against thyroglobulin.

Serum albumin in the elderly patient

Albumin values are somewhat lower in the elderly than in younger adults. Keating and his colleagues (1969) found that albumin levels fell by 0.095 g/litre per year of age over a wide range of ages whilst a similar decline (0.08 g/litre per year) is seen in elderly hospital patients (Northwick Park Hospital, $N = 680$).

Leask and his colleagues (1973) found the distribution of albumin values of well old people in a community survey in Scotland to be 41.1 ± 4.1 g/litre (mean ± standard deviation). In Harrow a group of 122 fit elderly subjects had somewhat higher values of 43.9 ± 3.1 g/litre whereas hospital patients in the same district have far lower values of 35.5 ± 5.5 g/litre (Northwick Park Hospital, $N = 605$). The albumin distributions for healthy old people in Harrow and their hospital counterparts are compared in Fig. 13. Albumin is not only 7.5 g/litre lower in the hospital patients but shows a wider spread of values, especially of lower values.

Low serum albumin in elderly patients most often seems to be due to the general effects of illness rather than to the specific effect of a particular disease. Albumin values tend to mirror the severity of

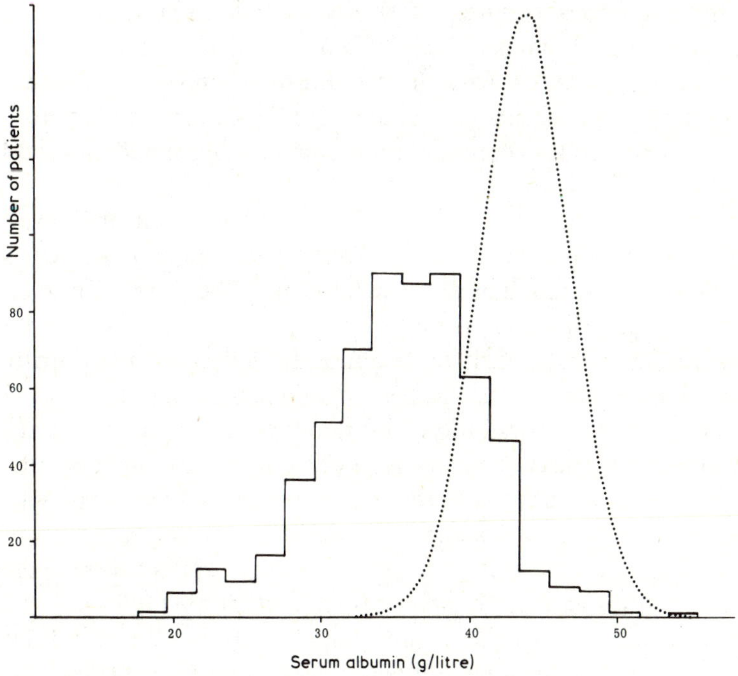

Fig. 13. Serum albumin results of 605 elderly Northwick Park Hospital patients (histogram) compared with those of healthy old subjects (normal curve).

illness, low values having a considerable adverse prognostic significance (Hodkinson, 1975). Albumin values of patients who are subsequently discharged are substantially higher than those of patients who die (36.4 ± 5.0 as compared with 33.7 ± 5.5 g/litre) though still considerably lower than the values in well elderly subjects (43.9 ± 3.1 g/litre).

A number of other plasma proteins which are synthesised by the liver are similarly affected. For example total iron binding capacity (a measure of the plasma concentration of transferrin) and T3-uptake (which roughly corresponds to TBG levels) also fall in ill old people (Hodkinson, 1975). Impairment of liver synthesis as a response to constitutional upset presumably underlies these phenomena.

It is uncertain how frequently such falls in albumin are due to the 'immediate response' or the 'delayed response' of the plasma proteins which have been described. The immediate response consists of a fall in albumin and rise in α-globulins and fibrinogen which return to normal within 10–14 days. Woodford-Williams and her colleagues showed these changes occured after myocardial infarction in the elderly (Woodford-Williams *et al.*, 1966). Similar responses may follow infections, inflammatory disease, burns, trauma or operation. The delayed response occurs in a wide variety of diseases and is more persistent. Hypoalbuminaemia is accompanied by raised α_2- and γ-globulins.

A low serum albumin is so often a feature of constitutional upset in old people that its development from more specific causes is overshadowed. Impaired hepatic synthesis may be a specific consequence of cirrhosis and other hepatocellular diseases. Hypoalbuminaemia may also result from abnormal losses as in nephrotic syndrome where there is heavy proteinuria. Protein losing enteropathy is another possibility. This may be seen in a wide variety of gastrointestinal diseases, ulcerative colitis giving particularly low albumin levels in elderly patients. Large pressure sores may give severe hypoalbuminaemia because of loss of serum, rapid improvement occurring once the sore is healed. Extensive stasis ulcers, large peritoneal effusions and extensive burns are further examples.

Hypoproteinaemia may also result from haemodilution as in fluid overload during intravenous infusions or in the syndrome of inappropriate ADH secretion. Conversely, haemoconcentration will elevate protein levels. Dehydration may result in a raised albumin value or, rather commonly in ill old people, cancel out a previously low one. Spurious elevation of albumin and other plasma proteins is also a consequence of excessive cuffing during venepuncture as already pointed out.

Serum albumin often gives a useful indication of the severity of illness in old people but rarely leads directly to a specific diagnosis. However, it is of additional value in the interpretation of other tests involving substances which are bound to albumin. This particularly applies to serum calcium which needs to be corrected for protein variation as will be more fully discussed in Chapter 6.

Serum globulin in the elderly patient

There is a gradual increase in serum globulins with age although this is less striking than the fall in albumin (Roberts, 1967). Leask and his colleagues (1973) found globulin values of well old people to be 30.5 ± 5.3 g/litre and values for 122 well old people in Harrow are very similar, 29.9 ± 3.9 g/litre. Elderly patients have somewhat higher values, Fig. 14 comparing those of 605 patients admitted to a

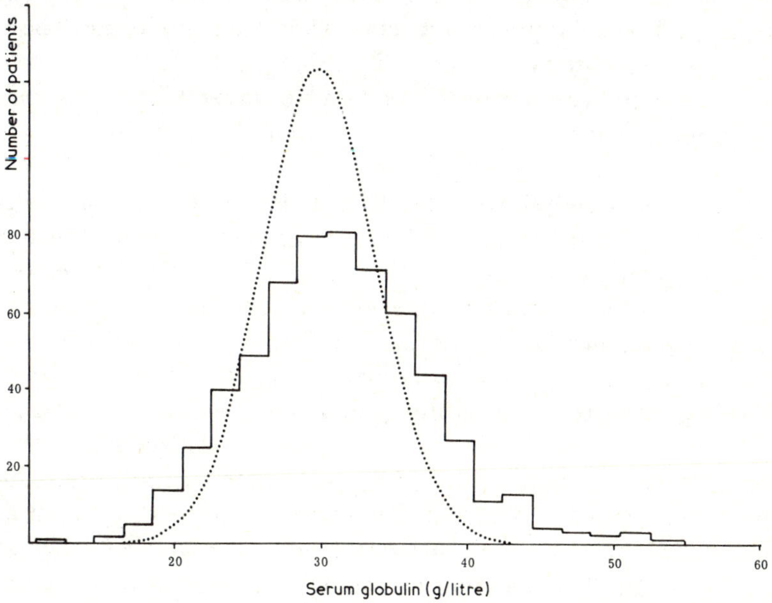

Fig. 14. Serum globulin results of 605 elderly Northwick Park Hospital patients (histogram) compared with those in healthy old people (normal curve).

geriatric department (Northwick Park Hospital) with the distribution in well old people.

Whilst measurement of serum globulin is justifiable as a part of the routine admission profile of tests ordered for the elderly patient, serum electrophoresis is better performed on a selective basis. It might reasonably be requested when serum globulin is 38 g/litre or more (37.8 g/litre being the upper 95% limit for the healthy elderly) or where there is a specific indication, for example when the diagnosis of myelomatosis is being considered. Immunoelectrophoresis may occasionally be of value, as for example in the detailed investigation of monoclonal gammopathies. The ranges for globulin fractions appropriate to the elderly have yet to be adequately established. However, Thomas and Powell (1971) found α_1-, α_2- and γ-globulins to be generally rather higher in elderly patients, and Maugham and her colleagues (1977) showed that IgA, IgM and most particularly IgG contributed to γ-globulin elevation.

When an elderly patient has a high globulin, electrophoresis usually shows this to be due to diffuse increase in one or more of the sub-groups. The α-globulins rise in both the immediate and delayed responses and when hypoalbuminaemia arises from any cause. The α- and β-lipoproteins are raised along with serum cholesterol in diabetes, obstructive jaundice, cirrhosis and severe myxoedema. Familial hyper-lipoproteinaemias are rarely seen in old age as longevity is generally reduced in these conditions.

Elevation of γ-globulins occurs in a wide variety of conditions, acute or chronic infections, neoplasms, autoimmune diseases such as rheumatoid arthritis and Hashimoto's disease and where tissue necrosis occurs. Such diffuse elevations of γ-globulin are thus of very limited diagnostic value, merely being a non-specific indication of disease.

In contrast, a monoclonal gammopathy, i.e. elevation of a single γ-globulin giving rise to a sharply defined peak on electrophoresis, is of far greater diagnostic significance though of infrequent occurrence. In elderly patients monoclonal gammopathy is most often due to myelomatosis and is then associated with Bence-Jones proteinuria in about 50% of cases. Alternatively it may be due to a benign gammopathy or Waldenström's macroglobulinaemia or may be associated with chronic lymphatic leukaemia, lymphosarcoma or carcinoma. Thomas and Powell (1971) should be consulted for further details of the differential diagnosis and more detailed investigation of monoclonal gammopathies in old age.

Elevation of total iron binding capacity due to the increase of transferrin in iron deficiency is a final example of globulin elevation due to a single protein species. Unfortunately it is not always seen in iron deficiency in elderly patients because of the contrary effects of constitutional upset which is so often present due to other causes.

The erythrocyte sedimentation rate

The ESR is a widely used test, upper limits of 5 mm/h for men and 15 mm/h for women generally being quoted. However, Boyd and Hoffbrand (1966) found that many ESRs above 20 or even 40 mm/h in elderly patients had no apparent cause whilst Sharland (1977) found a 95% range of 3–50 mm/h in apparently well elderly subjects who remained well at follow up and showed no significant change in serial ESRs.

The ESR principally reflects fibrinogen and globulin elevation but is also affected by changes in the haematocrit and albumin. Interpretation of a raised ESR is therefore uncertain and the test has little value other than as a general indicator of the presence of disease. Unfortunately, the very high upper limit in old age means that it is of very little practical value even in this limited role in the elderly patient. Serum albumin and globulin give more useful information and should replace the ESR in the investigation of the elderly patient.

CHAPTER FIVE

The Electrolytes, Urea and Tests of Renal Function

Hospital profiles of biochemical tests usually include the serum electrolytes and urea and quite often creatinine also. This chapter will mainly deal with interpretation of these commonly performed tests.

The electrolytes

Sodium, chloride, bicarbonate and potassium in that order are the main ions of the extracellular fluid and plasma and are commonly referred to collectively as the electrolytes. Intracellular fluid has a very different ionic composition, however, the main contributors in order of decreasing concentrations being potassium, phosphate, magnesium, sodium and bicarbonate. Thus some five sixths of the body sodium is in the extracellular fluid whilst approximately 97% of the body potassium is intracellular. An active ionic 'pump' using adenosine triphosphate as its energy source extrudes sodium from cells, potassium and hydrogen ions being taken into the cell in exchange.

The distributions of values for the serum electrolytes in the healthy elderly are closely similar to those in younger age groups. Table IV shows the values found for well elderly subjects in the community by Leask and his colleagues (1973).

Sodium

The serum sodium level is quite closely maintained in health. This homoeostasis depends on the kidney's ability to vary sodium and water resorption in the tubules so as to closely regulate their excretion in the urine.

Though the sodium levels in the healthy elderly subjects are the

TABLE IV

Distribution of serum electrolyte values in the healthy elderly (Leask et al., 1973)

Electrolyte	Mean (mmol/litre)	Standard deviation	95% range
Sodium	140.8	2.8	135–146
Potassium	4.4	0.4	3.6–5.2
Bicarbonate	25.1	2.8	20–31
Chloride	102.4	2.9	97–108

same as those in younger adults, this is not true of elderly patients. Their sodium values are set at a lower level and show a wider spread as is shown in Fig. 15 which compares the histogram of results from elderly patients admitted to a geriatric department with the

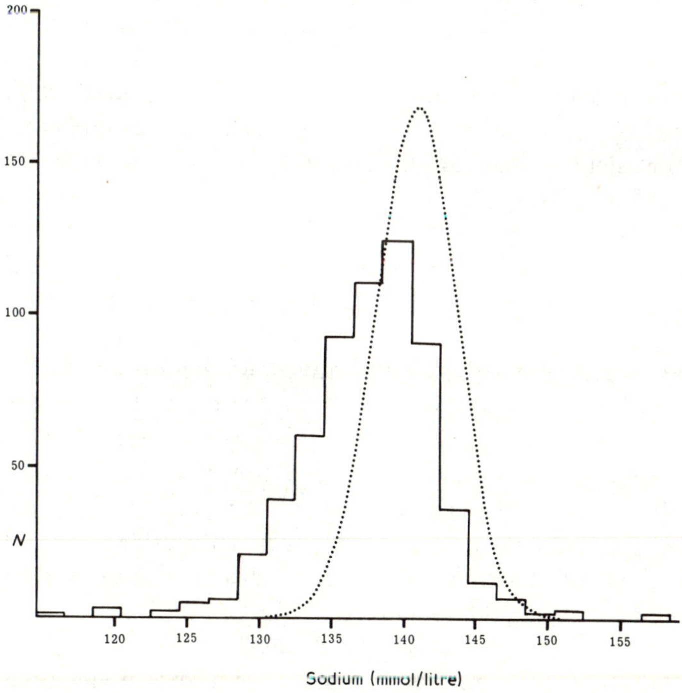

Fig. 15. Sodium results of 619 elderly Northwick Park Hospital patients (histogram) compared with those in healthy old people (normal curve).

distribution of values in the healthy elderly (shown as an idealised normal curve).

There are few high values. Only about 1% of the patients have values above the distribution for healthy subjects and these are chiefly due to severe dehydration. This, occurring in such situations as hyperosmolar diabetic coma, also results in very high values for total protein and haematocrit.

The many low serum sodium values occur for a variety of reasons. Abnormal sodium loss may be responsible and in the elderly patient hyponatraemia following diuretic therapy is by far the most common example and may occur even though there is considerable peripheral oedema. Abnormal renal loss of sodium may also occur in severe uraemia but sometimes occurs with less gross renal damage – the 'salt losing kidney'. Minor degrees of salt losing kidney may be fairly common in elderly patients with nephrosclerosis and give rise to hyponatraemia when salt intakes are low. Renal abnormal loss due to renal disease as opposed to the action of diuretics will usually be characterised by the presence of proteinuria and by elevation of blood urea.

Gastrointestinal losses of sodium may also result in hyponatraemia and this may be seen in vomiting, diarrhoea or in intestinal obstruction or ileus when large amounts of sodium may be sequestrated within the bowel.

Hyponatraemia may represent water retention rather than sodium loss – so called 'dilutional hyponatraemia' – and the two effects may sometimes be combined. Dilutional hyponatraemia appears to be especially common in elderly patients. Sometimes it may be due to therapeutic water overload during intravenous therapy or due to failure of the kidney to excrete a water load, for example when the renal blood flow is very low in such conditions as severe cardiac failure. More often it represents a specific failure of osmoregulation which may be precipitated by a wide variety of illnesses and conditions. The elderly seem to be especially vulnerable – indeed Mukherjee and his colleagues found evidence of defective osmoregulation even in well elderly subjects (Mukherjee *et al.*, 1973).

In its fully developed form, this type of dilutional hyponatraemia is called the syndrome of inappropriate anti-diuretic-hormone secretion or SIADH (Bartter and Schwartz, 1967). Hyponatraemia is accompanied by the secretion of a urine of high osmolality and with a significant sodium content. Administration of extra salt is unhelpful, the renal loss of sodium merely increases. Fluid restriction is the appropriate treatment. The syndrome is believed to be due to excessive secretion of ADH but alternatively an increased renal sensitivity to ADH may sometimes perhaps be responsible. SIADH may arise in a wide variety of diseases such as malignancies, respiratory infections, chronic respiratory disease, hypothyroidism, post-operatively or after trauma or head injury. Drugs may also be responsible, chlorpropamide quite often being to blame in elderly patients (Hodkinson, 1974). Although fully developed SIADH is quite common in elderly patients, minor degrees are probably very common in ill old people and perhaps account for the majority of instances of hyponatraemia.

Hyponatraemia coupled with a raised potassium occurs in Addison's disease but this is very uncommon in old age.

Potassium

Potassium values in elderly patients are also lower and more widely scattered than in the healthy elderly as can be seen in Fig. 16. Hyperkalaemia is infrequent provided that spuriously high values due to haemolysis or prolonged storage of blood specimens are avoided.

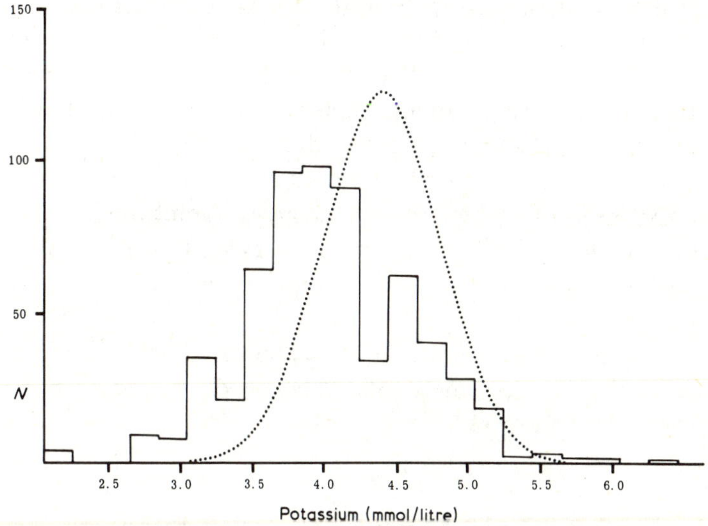

Fig. 16. Potassium results of 619 elderly Northwick Park Hospital patients (histogram) compared with those of healthy old people (normal curve).

The few cases seen either accompany severe renal failure or are due to overdosage of potassium supplements or to potassium retaining diuretics such as amiloride.

Hypokalaemia is frequent but is often of uncertain significance. Serum potassium represents less than 1% of the total body potassium and it is well recognised that serum potassium levels are a very poor guide to the overall potassium status. Ideally we should measure cellular rather than serum levels. Red blood cell potassium levels can be measured relatively easily but the analysis is not routinely available as yet. Serum potassium values have been shown to have no correlation with red blood cell potassium in elderly patients (Bahemuka and Hodkinson, 1976a) and so low serum potassium cannot be taken as evidence of potassium depletion without further corroboration. Potassium depletion due to poor dietary intake of potassium has been held to be the usual explanation of hypokalaemia in elderly patients, but red blood cell potassium evidence does not support this (Bahemuka and Hodkinson, 1976a, b).

Hypokalaemia may represent depletion of body potassium even though it is an unreliable guide. Usually abnormal losses are involved, either renal or gastrointestinal due to vomiting, diarrhoea or excessive mucus loss as in papillary carcinoma of the rectum. Increased renal loss associated with diuretics is often blamed. Its importance appears to have been much exaggerated, however (Davidson *et al.*, 1976). Indeed when elderly patients are given diuretics with potassium supplements their red cell potassium levels are greater than normal (Bahemuka and Hodkinson, 1976a).

Hypokalaemia may also occur when potassium is moved into cells. This occurs for example when insulin is given. Dilutional hypokalaemia may also occur during intravenous therapy. It is not known whether hypokalaemia in elderly patients may sometimes be related to SIADH.

Hypokalaemia may also result from alkalosis but potassium depletion itself gives rise to a metabolic alkalosis characterised by a raised bicarbonate and a low chloride.

Bicarbonate

Bicarbonate values in elderly patients have a wider spread than in healthy old people but a similar mean (25.6 ± 3.8 mmol/litre for 680 patients of the geriatric department, Northwick Park Hospital, compared with 25.1 ± 2.8 found by Leask *et al.*, 1973).

Low bicarbonate values are most often due to the metabolic acidosis of uraemia or diabetic ketosis but less commonly indicate lactic acidosis accompanying severe illnesses such as Gram negative septicaemia. Raised values may indicate respiratory acidosis when high pCO_2 and low pH are also found or may accompany the metabolic alkalosis of potassium depletion, alkali therapy, excessive vomiting or accompanying diuretic therapy.

Chloride

Chloride values in elderly patients are a little lower and more widely scattered than in well old people (99.3 ± 5.5 mmol/litre for 680 geriatric patients, Northwick Park Hospital, compared with 102.4 ± 2.9 found by Leask *et al.*, 1973). Chloride values are strongly correlated with those of sodium and these changes are largely a reflection of lower and more variable sodium levels in ill old people. Chloride tends also to be inversely correlated with bicarbonate and its greater variability as a result of impairments of acid–base balance must also contribute to the variation in chloride.

Chloride values are generally of limited usefulness in the routine investigation of elderly patients and are often excluded from biochemical profiles now.

Osmolality

Reference has already been made to the impaired osmotic homoeostasis in old age. Osmotic activity of a solution, its osmolality, depends on the number of solute particles (molecules or ions) in a given mass of solvent. In other words it depends on molalities rather than the molarities (solute concentration per unit volume of solution) but it is the latter that are usually measured biochemically because of the conveniences of volumetric analyses.

Serum osmolality can be measured fairly easily by the freezing point depression method. Alternatively, it can be estimated approximately from molarities of the main serum constituents. Calculation can be simplified by ignoring the anions and including only the main cations as the two equal each other. A variety of formulae have been proposed although many are theoretically rather than practically based. In elderly patients the following formula predicts osmolality with a standard error of the estimate of only 1.4% (Hodkinson, 1975).

Serum osmolality (mosm/litre) = 38.64 + (sodium mmol/litre x 1.545) + (urea mmol/litre x 1.080) + (glucose mmol/litre x 0.763) + (total protein g/litre x 0.253)

Sodium is by far the most important contributor to calculated osmolality. Correspondingly, measured osmolality usually gives little information which is not given by serum sodium. The measurement or calculation of osmolality appear to have no useful part to play in the routine investigation of elderly patients. Simultaneous serum and urine osmolalities are of great value in confirming SIADH as the cause of hyponatraemia, however.

The normal range for osmolality in elderly patients is 260–309 mosm/litre (Hodkinson, 1975) as compared with a range of 281–297 mosm/litre in healthy adults (Diem and Lentner, 1970). Low values equate with hyponatraemia, the causes of which have already been considered. High values are commonly due to uraemia, hyperglycaemia or dehydration.

Tests of renal function

The kidney has a central role in the maintenance of the constancy of the 'milieu intérieur' by the excretion of waste products, retention of useful metabolites and regulation of acid–base balance. The assessment of renal function is therefore of great importance generally but especially in elderly patients where disturbances are particularly common. Furthermore, knowledge of the adequacy of renal function is very important where treatment regimes involve drugs excreted by the kidney. Renal impairment also affects biochemical test results as already mentioned in Chapter 3.

The tests of renal function commonly used – the measurement of serum urea and creatinine and tests for proteinuria – are very valuable in assessing renal integrity but do not usually give any aetiological diagnostic information. This usually depends on clinical information and special diagnostic tests which are outside the scope of the present chapter.

Glomerular filtration rate

The glomerular filtration rate (GFR) is roughly reflected by both serum urea and creatinine. The accurate determination of GFR based on inulin clearance is essentially a research rather than a clinical procedure. However, two reasonably accurate measures of GFR are available in clinical practice. Creatinine clearance utilising endogenous creatinine but involving an accurate timed urine collection as well as simultaneous blood specimens has been widely used. However, the need for careful urine collections often renders it totally impracticable in the elderly patient. Fortunately, a far more practical method is now available using radioactive-chromium-labelled EDTA (^{51}Cr-EDTA). Blood specimens but not urine specimens are required so that the method is eminently suitable for use in the elderly (Denham *et al.*, 1975). GFR measured by either of these methods is usually expressed in terms of standard body surface area of 1.73 m^2, the patient's surface area being calculated from height and weight. This correction allows comparisons of GFR to be made irrespective of variations in body size.

Both urea and creatinine levels in the serum are largely dependent on GFR. However, both are also influenced by other factors. Urea is the main end-product of protein breakdown. The rate of urea production can therefore vary considerably, being increased when dietary protein intake is high or when protein catabolism is increased in disease. In contrast, creatinine is unrelated to protein intake as it is produced by the breakdown of muscle proteins. Instead, it is related to muscle mass and hence to body weight. Creatinine production will also be higher if muscle destruction occurs.

In younger patients, creatinine is generally considered to be a more reliable guide to GFR than urea. However, this is not the case in elderly patients where urea is marginally superior to creatinine even when the latter is corrected for variations in body weight (Denham *et al.*, 1975).

Renal function in old age

In common with most bodily functions, renal function shows a steady deterioration with age. The number of functioning nephrons falls and GFR progressively declines (see review by Sourander, 1973). Correspondingly, ranges for urea and creatinine rise with age throughout adult life (Roberts, 1967; Keating *et al.*, 1969). Creatinine changes may be partly annulled by decreasing muscle mass with age, however.

Urea

Leask and his colleagues (1973) found a range for urea of 3.9–9.9 mmol/litre (23–59 mg/100 ml) in the healthy elderly, women having slightly lower values than men. This range may be compared with that of 3.5–7.2 mmol/litre (21–43 mg/100 ml) for adults under 65 (Roberts, 1967). Figure 17 compares the urea values of elderly patients with those of healthy elderly subjects, the distributions being log-normal. The values of the patients are far more widely spread and generally far higher, indicating the considerable prevalence of renal impairment due to pre-renal, renal and post-renal pathologies. However, there is also an appreciable number of abnormally low urea values in the hospital patients. These may be dilutional due to fluid overload or SIADH or due to low dietary intakes of protein.

Urea values of elderly patients give a reasonably good indication of GFR despite the disturbing effects of dietary intake of protein and abnormal catabolism. Figure 18 shows the correlation between GFR and urea in elderly patients (Denham *et al.*, 1975). Log scales have been used as both GFR and urea results are log-normally distributed. The relationship is linear when the log transformations are used. The

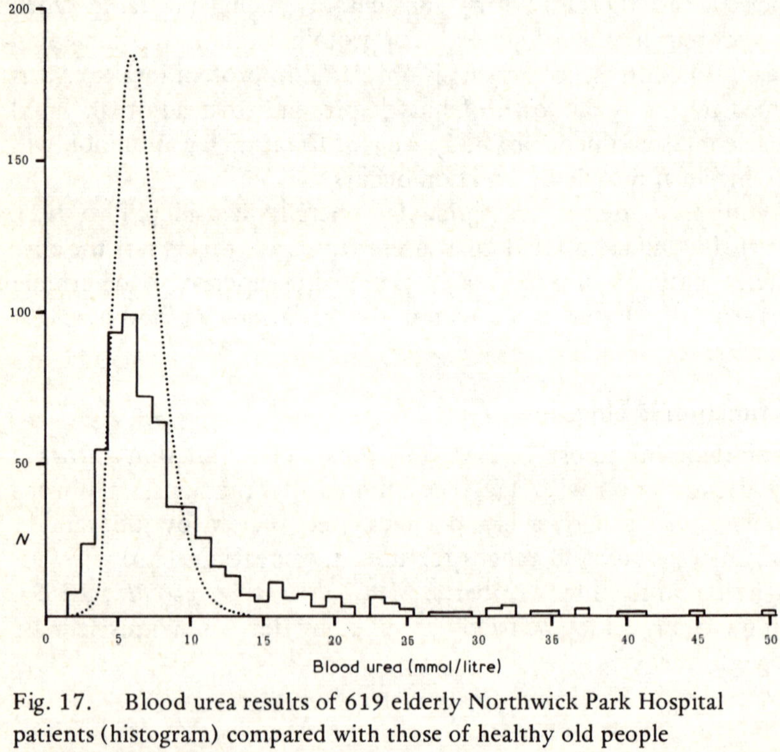

Fig. 17. Blood urea results of 619 elderly Northwick Park Hospital patients (histogram) compared with those of healthy old people (log-normal curve).

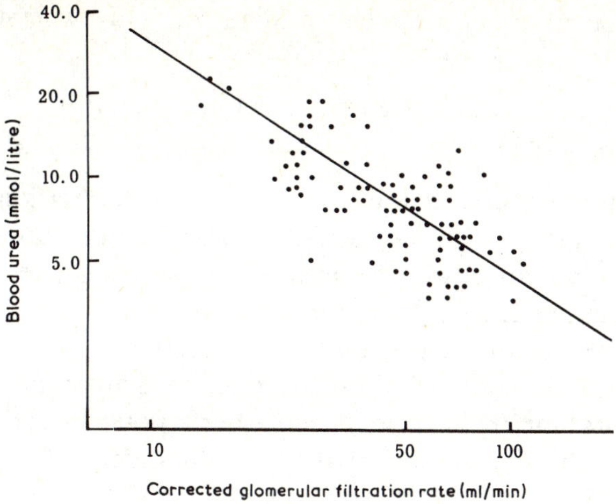

Fig. 18. Correlation of blood urea and corrected GFR results in elderly patients (after Denham *et al.*, 1975).

correlation coefficient is 0.71 – this implies that 55% of the variance of urea is accounted for by variations in GFR, the remainder representing variations due to within and between person, analytical and dietary protein intake factors. Prediction of GFR might be improved if allowance could be made for protein intake but this is impracticable in routine clinical practice. However, we should remember that grossly unusual protein intakes, for example the use of a low protein diet in chronic renal failure, will invalidate the prediction of GFR from urea.

Creatinine

Creatinine is also log-normally distributed, the normal range in healthy old people being 52–159 μmol/litre (0.6–1.8 mg/100 ml) (Leask *et al.*, 1973). Hospital patients have a higher range of 51–222 μmol/litre (0.6–2.5 mg/100 ml) (Bahemuka *et al.*, 1974). Creatinine varies with body weight and elderly patients show a very wide range of weights – tiny elderly ladies may have apparently normal creatinine values even when their renal function is considerably impaired. Creatinine values need to be corrected for body weight before they can be safely interpreted and such correction abolishes the apparent sex difference in creatinine results. A nomogram is available which allows prediction of GFR from creatinine, age and weight (Kampmann *et al.*, 1974). In elderly patients the formula of Denham and his colleagues may be used to correct creatinine for body weight (Denham *et al.*, 1975):

$$\log \text{corrected creatinine } \mu\text{mol/litre} = \log \text{creatinine } \mu\text{mol/litre} - (\text{weight kg} \times 0.00247)$$

These authors found that corrected creatinine gives a correlation with GFR which is marginally less strong than that for urea ($r = 0.73$). Creatinine has no advantage over urea as a single test estimating GFR in elderly patients therefore and is less convenient in practice because of the need to apply a weight correction. Furthermore, though creatinine is said to be less affected by dehydration than urea, this appears not to be true in the elderly patient, both rising in parallel in the dehydrated patient (Sourander, 1973; Bahemuka *et al.*, 1974).

Combined use of urea and creatinine

Though there is little to choose between urea and creatinine individually, GFR can be predicted somewhat more reliably in the elderly if the two results are combined (Denham *et al.*, 1975). Figure 19 shows a nomogram based on the formula given by these authors which allows GFR to be estimated from urea, creatinine and body weight. The 95% confidence limits for the resulting estimate of GFR are 52–193% of actual GFR, a modest improvement on those of 49–205% for GFR estimated from urea alone.

Urea and creatinine values are usually strongly correlated (Bahemuka *et al.*, 1974) and the line on the nomogram which joins the point on the line C – C (which represents corrected creatinine) with the urea value will usually be roughly horizontal. When it departs markedly from the horizontal it suggests that there may be some unusual disturbance of one or other test. If the urea value is unexpectedly low this may indicate a very low protein intake or the occurrence of water overload as in SIADH. If urea is unexpectedly high this may be due to increased protein catabolism resulting from illness, fever, trauma or the anti-anabolic effects of corticosteroid therapy. A proportionately high urea is not a reliable guide to the presence of dehydration in elderly patients where the creatinine and urea levels tend to rise in parallel. Other tests are of more value in making the distinction between dehydration and renal failure. Thus the serum proteins and packed cell volume are usually raised in dehydration whilst a high phosphate and low bicarbonate support the diagnosis of renal failure.

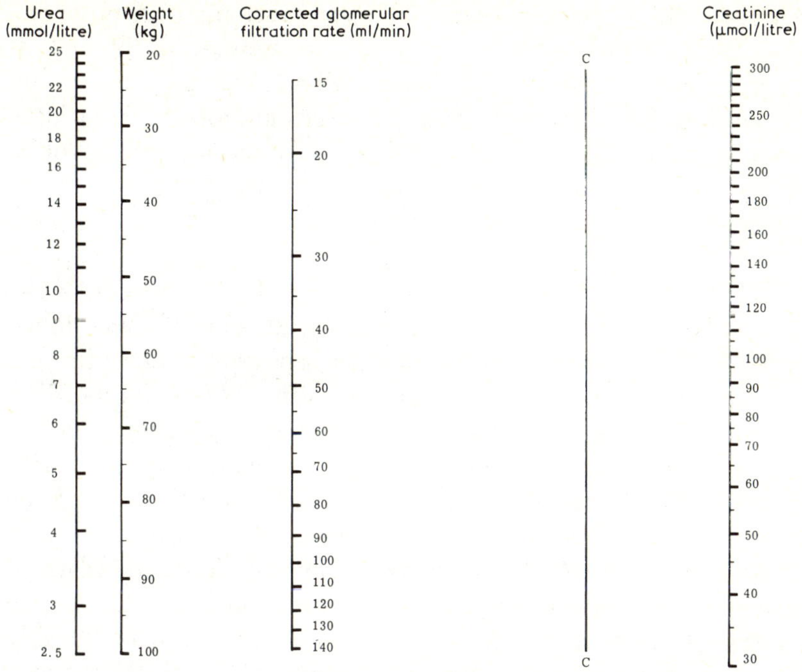

Fig. 19. Nomogram for estimation of corrected GFR of elderly patients (after Denham *et al.*, 1975). (*Method*: join serum creatinine to weight by a line crossing C – C. Join point of intersection to urea by line indicating result on GFR scale.)

Proteinuria

Small amounts of protein are lost in the urine in health but are insufficient to give positive results with the tests for proteinuria which are usually employed such as the boiling, salicyl-sulphonic acid or stick tests. These tests are, however, sufficiently sensitive to give positive results if the urine is contaminated by vaginal secretions or if it contains blood or pus. Careful urine collection and microscopy of the urine can eliminate these possibilities. Given these exclusions, proteinuria usually indicates structural disease of the kidney or urinary tract.

Glomerular disease is most often responsible, pyelonephritis, nephrosclerosis and diabetic nephropathy being major possibilities in the elderly patient. Renal papillary necrosis sometimes occurs. Acute glomerulonephritis and the nephrotic syndrome are both very rare in old age, however.

Alternatively, proteinuria may indicate myelomatosis, not renal disease. Bence-Jones proteinuria occurs in approximately half of cases and can be recognised by precipitation followed by re-solution in the boiling test or, more reliably, by electrophoresis of the urinary protein. Serum electrophoresis will show a monoclonal globulin band.

Uric acid

Uric acid may be briefly mentioned as a test which is very much affected by impairment of renal function. The normal range for uric acid is somewhat higher in the healthy elderly, 148–461 μmol/litre (2.5–7.7 mg/100 ml) (Leask *et al.*, 1973), than in younger adults. In elderly patients uric acid values are quite strongly correlated with urea and with creatinine ($r = 0.58$ and 0.53 respectively) and the generally poorer renal function results in an elevated 95% range of 162–682 μmol/litre (2.7–11.4 mg/100 ml) for uric acid (data of Denham *et al.*, 1975). The test is of very limited value in the diagnosis of gout in elderly patients as it can only be reliably interpreted in those with good renal function. Uric acid is not a useful test to include in the routine profile for elderly patients, therefore.

CHAPTER SIX

Calcium, Phosphorus and Alkaline Phosphate

This chapter deals with the use of serum calcium, phosphorus and alkaline phosphatase in the diagnosis of metabolic diseases of bone. Amongst these, osteomalacia is of particular importance both because of its amenability to effective treatment and its relatively high prevalence in the elderly (Anderson *et al.*, 1966).

The diagnosis of osteomalacia is the principal reason for the inclusion of these tests in biochemical profiles of elderly patients but they also serve to identify hypercalcaemias from a variety of metabolic causes. These have a considerable prevalence (Grero and Hodkinson, 1977), indeed in southern England they are more common than osteomalacia. This may not be the case in the north of England or in Scotland where higher prevalences of osteomalacia appear to be encountered (Anderson *et al.*, 1966; Hodkinson, 1975).

Although used in the diagnosis of metabolic bone diseases, calcium, phosphate and alkaline phosphatase are more often altered by other causes in the elderly patient.

Calcium

Calcium tends to fall with age in men but to rise with age in women (Roberts, 1967; Keating *et al.*, 1969; Reed *et al.*, 1972). Leask and his colleagues (1973) found values of 2.39 ± 0.10 mmol/litre (9.55 ± 0.42 mg/100 ml) for men and 2.43 ± 0.13 mmol/litre (9.72 ± 0.53 mg/100 ml) for women in their healthy elderly subjects. Very similar values, 2.39 ± 0.10 mmol/litre (9.56 ± 0.40 mg/100 ml) for men and 2.44 ± 0.09 mmol/litre (9.76 ± 0.36 mg/100 ml) for women, were found in 122 well elderly subjects in Harrow.

The difference between the sexes appears to be due to a sex-hormone effect. Sex hormones antagonise the effects of parathormone on bone. The lower hormone levels of post-menopausal women favour parathormone dependent osteoclastic resorption of bone and the development of higher serum calcium and phosphate levels (Young *el al.*, 1968; Riggs *et al.*, 1969).

Elderly patients have generally lower calcium levels than those of healthy elderly subjects. Values in elderly patients admitted to a geriatric department were 2.32 ± 0.17 mmol/litre (9.28 ± 0.68 mg/100 ml) (Hodkinson, 1973b). However, the lower values largely relate to the lower albumin values of ill old people, calcium being bound to the serum proteins and most particularly albumin to a considerable degree. As the body homoeostatic mechanisms are concerned with the levels of ionised (unbound) calcium, lowered protein levels are associated with lowered calcium values without there being any disturbance of calcium metabolism.

It is thus of special importance to correct calcium values for protein variation in elderly patients where proteins are so often affected in illness. An appropriate correction formula derived from the results of 1447 elderly patients (Hodkinson, 1975) is:

corrected calcium mmol/litre = calcium mmol/litre – (albumin g/litre x 0.01833) – (globulin g/litre x 0.00639)

This correction can easily be made using a pocket calculator but alternatively the nomogram shown in Fig. 20 may be used.

Corrected calcium values in the elderly

Corrected calcium values for 122 healthy elderly subjects in Harrow were 1.40 ± 0.10 in men and 1.45 ± 0.09 mmol/litre in women when the above correction was used. Elderly patients have a mean corrected calcium of 1.44 mmol/litre but show a considerably wider spread of values as is shown for women in Fig. 21. Only a minority of the excess of both high and low values represents osteomalacia and metabolic bone disease. Many of the extreme values are due to other disturbing effects. In particular, many are associated with elevation of blood urea as is shown in Fig. 22. Hypocalcaemia is a common feature of chronic renal failure whilst a small number of patients develop renal osteomalacia. Hypercalcaemia may result in renal failure. However, in elderly patients, the association of high calcium values with a high urea is more often seen in severe dehydration. Hypercalcaemia occurs transiently in dehydrated elderly patients quite often (Hodkinson, 1973b). This is probably because of a severe fall in calcium excretion due to the very low GFR.

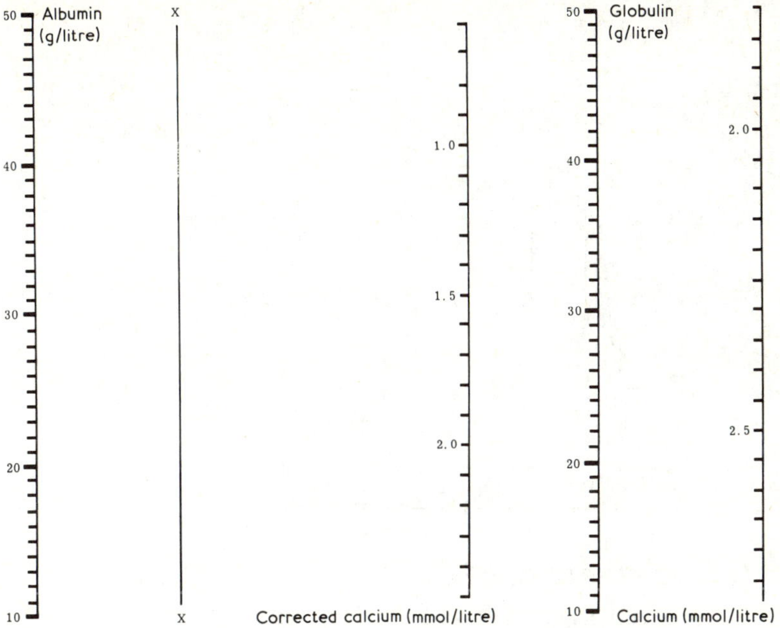

Fig. 20. Nomogram for correction of serum calcium. (*Method*: join albumin and globulin results by line crossing X – X. Join point of intersection to calcium result by line indicating result on corrected calcium scale.)

Treatment with thiazide diuretics or frusemide may favour its development as these drugs inhibit calcium excretion (Middler *et al*., 1973).

Vitamin D overdosage is sometimes responsible for hypercalcaemia and often is due to self medication with vitamin pills.

Hypocalcaemia is quite often seen as a consequence of treatment with sex hormones, anabolic steriods or carbenoxolone. Stilboestrol can give considerable hypocalcaemia and hypophosphatasia; osteomalacia may be wrongly suspected if the clinician is unaware of these effects. Hypocalcaemia also results from glucose adminstration as in the glucose tolerance test, the intravenous injection of glucose or during glucose or glucose/saline infusions.

Phosphate

The changes in phosphate values with age closely resemble those for calcium as both relate to sex-hormone changes. Thus phosphate values fall with age in men but rise slightly in women. There is quite

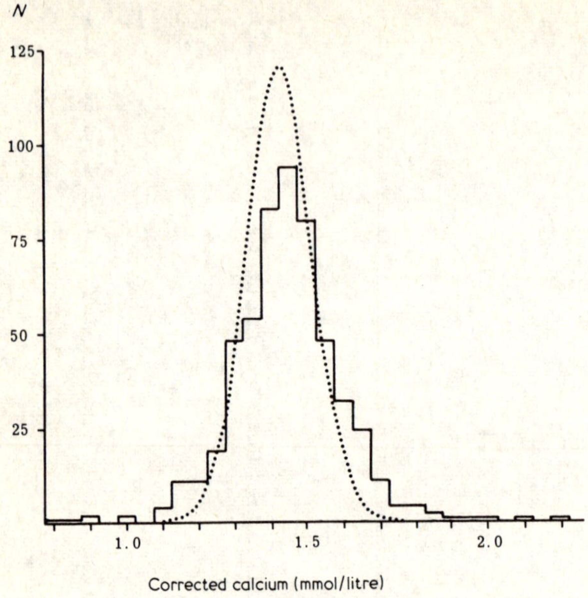

Fig. 21. Corrected calcium results of 540 elderly female Northwick Park Hospital patients (histogram) compared with those of healthy old women (normal curve).

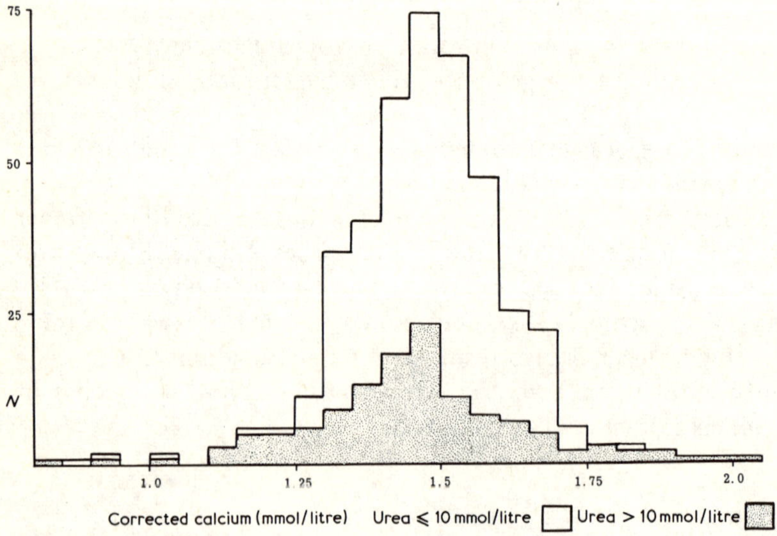

Fig. 22. Corrected calcium results of 540 elderly female Northwick Park Hospital patients showing their relationship to elevation of blood urea (shaded areas = patients with blood urea greater than 10 mmol/litre).

a striking difference between values of elderly men and women, far greater than the sex difference for calcium. Leask and his colleagues (1973) found values of 0.99 ± 0.16 (3.06 ± 0.49 mg/100 ml) and 1.09 ± 0.19 mmol/litre (3.39 ± 0.60 mg/100 ml) for healthy elderly men and women respectively. The values for 122 well old people in Harrow were similar – 0.94 ± 0.15 (2.9 ± 0.46 mg/100 ml) and 1.09 ± 0.12 mmol/litre (3.38 ± 0.37 mg/100 ml) for men and women respectively – but show tighter distributions. This probably relates to the marked diurnal variation for phosphate (Fig. 1), the Harrow series blood specimens all being taken mid-morning whereas those in the study of Leask and his colleagues were taken either in the morning or afternoon.

Phosphate results in elderly patients show considerable disturbance when compared with those of well old people (Fig. 23). There is marked skewing to higher values which is almost entirely attributable to renal impairment (Hodkinson, 1973a). However, a small number of high values represent other causes such as diabetic coma.

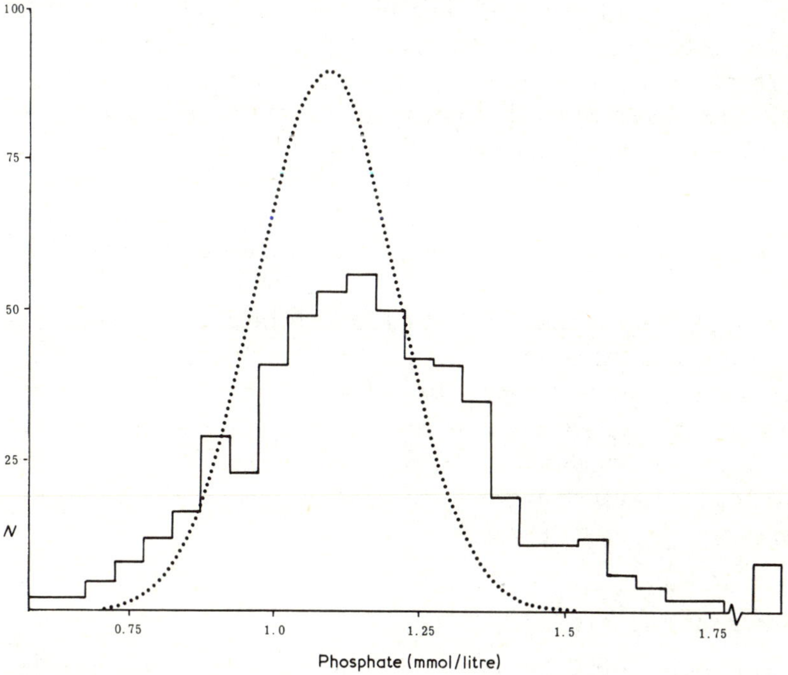

Fig. 23. Serum phosphate results of 540 elderly female Northwick Park Hospital patients (histogram) compared with those in healthy old women (normal curve).

Some of the low phosphate values are due to osteomalacia, hyperparathyroidism or other metabolic hypercalcaemias, but many are due to a variety of other causes. Thus, like calcium, phosphate is lowered by treatment with sex hormones, anabolic steroids, and by glucose administration and also falls after meals or with insulin (Betro and Pain, 1972). Antacids such as aluminium hydroxide also lower phosphate because phosphate absorption from the gut is hampered by the formation of insoluble phosphate salts.

Alkaline phosphatase

Alkaline phosphatase values tend to rise with age (Roberts, 1967; Keating *et al.*, 1969; Reed *et al.*, 1972) and higher ranges than those for younger adults have been reported for apparently healthy subjects (Hobson and Jordan, 1959; Sharland, 1972; Leask *et al.*, 1973). However, it is not clear whether these changes might not be due to occult diseases such as Paget's disease rather than to a genuine age change. Detailed studies of hospital patients indiate that if disease which might affect alkaline phosphatase results is carefully excluded, alkaline phosphatase is no higher in old age than in earlier age groups (Hazell and Ortiz, 1966; Hodkinson and McPherson, 1973). The latter authors found the range in old age to be 3.6–14.5 kAU, values being log-normally distributed.

However, many elderly patients show elevated alkaline phosphatase values, as is shown in Fig. 24. Elevation occurs for many reasons, osteomalacia being responsible relatively infrequently; liver disease, fractures, malignant deposits in bone, Paget's disease, periostitis secondary to chronic leg ulceration, rheumatoid arthritis and gastric surgery all being more common (Hodkinson and McPherson, 1973). Other less common causes of elevation of alkaline phosphatase in the elderly patient include senile cardiac amyloidosis (Hodkinson and Pomerance, 1974), ulcerative colitis, steatorrhoea (Dent *et al.*, 1968) and giant cell arteritis (Hall and Hargreaves, 1972).

Screening for osteomalacia

The difficulties in interpreting calcium, phosphate and alkaline phosphatase results as screening tests for osteomalacia in elderly patients are very considerable. We have already shown that all three

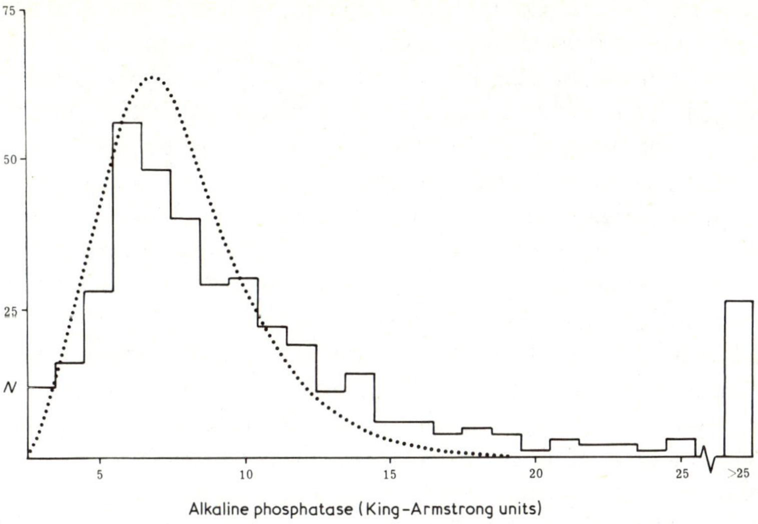

Fig. 24. Serum alkaline phosphatase results in 388 elderly female Northwick Park Hospital patients (histogram) compared with those in healthy old women (log-normal curve).

tests have low specificity. In particular, impairment of renal function greatly complicates the interpretation of calcium and phosphate results. Because of this we will first consider interpretation in those with reasonably good renal function (urea 10 mmol/litre or less) and discuss the diagnostic problems in patients with poorer renal function subsequently.

We have already seen (Chapter 2) that screening is far more difficult when the prevalence rate of the disease is low. What are the likely prevalence rates of osteomalacia in elderly patients? We have already indicated that they may vary with climate, higher rates having been reported from Scotland (Anderson *et al.*, 1966) than from southern England (Hodkinson, 1975). There is undeniably a marked difference between the sexes, osteomalacia being predominantly a disease of the elderly female (Chalmers *et al.*, 1967). Experience at Northwick Park Hospital indicates that the prevalence in elderly men is rather less than $\frac{1}{2}$%, screening being of dubious value except in post-gastrectomy patients or those on enzyme inducing drugs such as the barbiturates. The prevalence in elderly women without gastrectomy is of the order of 2% whilst after gastrectomy it is perhaps as high as 20%.

Let us examine the problems of screening in the elderly female patient without serious renal impairment, assuming a prevalence of 2%. Table V summarises the biochemical findings of elderly female patients with and without histological evidence of osteomalacia (Northwick Park Hospital). All had a blood urea of 10 mmol/litre or less and had no other disease or therapy likely to disturb their calcium, phosphorus or alkaline phosphatase values. It is apparent that none of the three tests adequately distinguishes between the normal and osteomalacic patients. Calcium gives the least unsatisfactory separation, 2.1 pooled standard deviation units, phosphate and alkaline phosphatase being closely similar (1.6 S.D.). Application of the method of calculation of posterior probabilities shown in Chapter 2 given a prior probability of 2% for osteomalacia shows that only 33% of calcium values of osteomalacic patients will lead to a posterior probability of 0.5 or better (i.e. will lie beyond the critical limit). Phosphate and alkaline phosphatase do even less well with only 30% of results outside the critical limit.

It is essential that we combine the test results in some way so as to improve their discriminatory power. The calcium–phosphorus product has been much used although its theoretical basis is far from secure (see Nordin (1973) for discussion of this). In practice it discriminates less well than a discriminant function using calcium and phosphate values and far less well than one combining all three biochemical tests. However, it does rather better than any of the single tests, the percentage of results outside the critical limit rising

TABLE V

Comparison of biochemical findings in elderly women with and without histological evidence of osteomalacia whose urea values did not exceed 10 mmol/litre

Test	Mean ± S.D. pts with normal bone histology $N = 27$	Mean ± S.D. patients with histological osteomalacia $N = 18$
Corrected calcium (mmol/litre)	1.44 ± 0.10	1.21 ± 0.12
Phosphate (mmol/litre)	1.06 ± 0.13	0.80 ± 0.21
$\log_e$ alkaline phosphatase (kAU)	2.40 ± 0.39	3.18 ± 0.64

to 43%. The discriminant function based on all three tests gives 50% of results outside the critical limit. The formula for calculation of the function is:

(corrected calcium mmol/litre x 6.115) + (phosphate mmol/litre x 2.593) – ($\log_e$ alkaline phosphatase kAU x 0.9605) – 9.216,

the value of the function then being 0.0 ± 1.0 in normal patients and 2.80 ± 1.0 in patients with osteomalacia. Though this can be readily calculated and a posterior probability estimated if a programmable calculator is available, it is not otherwise practicable in the clinical situation unless a nomogram is used. The nomogram in Fig. 25 allows it to be used with any prior probability, the posterior probabilities shown being those originating from a prior probability of 0.5 (1 : 1). The posterior probability for other prior probabilities is obtained by simple multiplication of the nomogram probability and the prior probability, both being expressed as 'odds'. For example if the

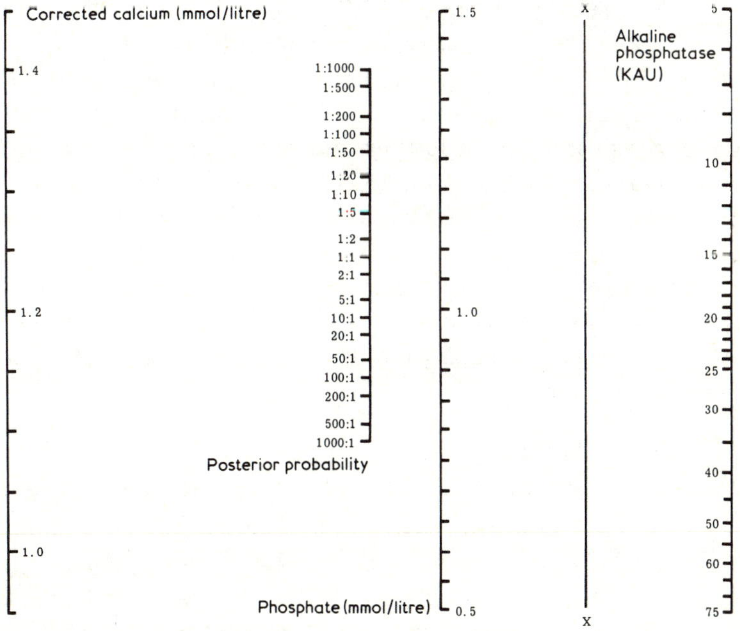

Fig. 25. Nomogram for estimation of posterior probability of osteomalacia in an elderly female patient when prior probability is 0.5. (*Method*: join alkaline phosphatase and phosphate results by line crossing X – X. Join point of intersection to corrected calcium result by a line indicating posterior probability on scale.)

nomogram gives a posterior probability of 1 : 100 and the appropriate prior probability is 2 : 1 (i.e. 0.67 for osteomalacia), the true posterior probability is (1 x 2) : (100 x 1) or 1 : 50.

The discriminant function, though somewhat difficult to use, does allow the maximum diagnostic information to be extracted from the three test results. If it is not used, the simple pattern recognition approach familiar to clinicians is far from satisfactory. This is based on the expectation that the osteomalacic patient will have calcium and phosphate results below and alkaline phosphatase above the appropriate normal ranges. However, in the 18 osteomalacic patients considered in Table V, only 6 (33%) had all three tests abnormal, 56% had two abnormalities whilst 11% had only a single abnormal test.

The discriminatory power of calcium, phosphate and alkaline phosphatase even when combined as a discriminant function is far from satisfactory. The posterior probability must often fail to give a clear cut answer as to whether or not the patient has osteomalacia. We have already seen in Chapter 2 that in such circumstances a repeat set of tests is only of marginal assistance. Ideally we need to turn to an independent test. A radiological search for Looser's zones is very valuable for, if they are found, they are virtually pathognomonic. Iliac crest bone biopsy is the other definitive method (though not without its own interpretational difficulties) but is a somewhat unpleasant procedure from the patient's point of view and certainly should only be performed when the probability of osteomalacia is reasonably high. An independent biochemical test with good discriminatory power would thus be of great value. In consequence there is considerable interest in the measurement of circulating vitamin D levels.

Serum 25-hydroxy-cholecalciferol

The measurement of 25-hydroxy-cholecalciferol (25-HCC) has become a practical possibility in recent years although the test is not yet generally available. It may well become so in the near future. It is the main circulating form of vitamin D but not the most active metabolite – this being 1,25-dihydroxy-cholecalciferol (1,25-DHCC) synthesised in the kidney from 25-HCC.

Unfortunately, though 25-HCC levels are usually very low in osteomalacia, its discriminant value in the elderly is yet to be assessed and the appropriate normal range to be determined. Many institutionalised elderly people have very low 25-HCC values without other clear evidence of osteomalacia (Corless *et al.*, 1975). Furthermore, 25-HCC may not be a truly independent test, Brown and his colleagues (1976) having found a positive correlation between 25-HCC and serum calcium values in elderly patients. Results may also be distorted by serum protein changes, 25-HCC being strongly protein bound, mainly to a specific α_1-globulin carrier protein (Peterson, 1971). It is thus likely that elderly patients who have severe constitutional upset may have spuriously low 25-HCC values in the same way that low serum iron, T4 or calcium values may occur. Further work is needed to determine whether or not 25-HCC measurements will be of practical help in screening elderly patients for osteomalacia. Measurement of 1,25-DHCC might be of greater value should it become clinically available as it is the active form of the vitamin. Certainly it should be more reliable in patients with renal impairment where failure to convert 25-HCC to 1,25-DHCC might be responsible for osteomalacia.

Diagnosis of osteomalacia in the presence of renal failure

The diagnosis of osteomalacia in elderly patients with substantial renal impairment is far more difficult than in those with good renal function. This is because phosphate retention is usual so that hypophosphataemia is no longer a useful guide to the diagnosis and because hypocalcaemia quite commonly occurs in the absence of osteomalacia. Hypocalcaemia may thus be a result of acidosis which tends to lower calcium binding to proteins so that corrected calcium values suggest a lower ionised calcium than is the case. Furthermore, phosphate elevation tends to produce lowering of serum calcium. In acute renal failure this is probably because bone mineralisation is stimulated but in chronic renal failure tissue deposition of calcium phosphate is perhaps responsible (Nordin, 1973).

In time, hypocalcaemia tends to lead to bone disease. Secondary hyperparathyroidism is a usual response to hypocalcaemia, increased resorption of bone partially compensating for the low serum calcium levels but at the cost of development of parathyroid bone disease if the process is sufficiently severe and prolonged. Some patients develop osteomalacia characterised by relative resistance to vitamin D therapy, presumably because of the impaired ability to convert 25-HCC to 1,25-DHCC.

In either of these two cases, alkaline phosphatase elevation usually occurs and it is this test which, despite its low specificity, provides the best guide to the development of bone disease in the elderly patient with hypocalcaemia and renal failure. Phosphate values are unhelpful but patients with bone disease tend to have particularly low calcium values and more marked lowering of the plasma bicarbonate (Nordin, 1973).

The biochemical findings are, however, a generally poor guide to the presence of bone disease and one must rely more on the radiological finding of Looser's zones as an indication of osteomalacia or of the erosive or cystic changes indicating hyperparathyroidism, or on iliac crest bone biopsy histology. Biopsy is not always justifiable in severely ill elderly patients with renal failure whose general prognosis is poor and, indeed, a full diagnosis may be academic in such circumstances.

Hypoparathyroidism

Hypoparathyroidism as a cause of hypocalcaemia is very uncommon in elderly patients. Such few cases as are seen most usually result from parathyroid damage at thryroid surgery. Typically the hypocalcaemia is accompanied by elevation of phosphate whilst alkaline phosphatase is unaffected.

Hypercalcaemia

Leaving aside the transient hypercalcaemias that may be seen in severely dehydrated patients and single high calcium values attributable to biological and analytical variation, persistent hypercalcaemia is an important and quite common finding in elderly patients. Grero and Hodkinson (1977) found it in $1\frac{1}{3}$% of patients admitted to a geriatric department. Occasionally, hypercalcaemia may be due to vitamin D intoxication either because of over-enthusiastic treatment or to self medication with vitamin preparations. Otherwise, hypercalcaemia in the elderly patient may be associated with a number of serious diseases. As with hypocalcaemia, diagnosis is far less certain when renal failure is present and renal failure may result from hypercalcaemia. The main causes of hypercalcaemia in elderly patients with good renal function were found to be (Grero and Hodkinson, 1977):

(1) osteolytic bone metastases (37%);
(2) primary hyperparathyroidism (27%);
(3) carcinoma of bronchus without bone metastases (23%);
(4) lymphoma without bone metastases (10%).

They found quite marked biochemical differences between these four groups which, together with clinical and radiological findings, will allow ready classification of many patients. Hypercalcaemia due to bone metastases was typically associated with a high alkaline phosphatase, low albumin, raised globulin and normal phosphate. Primary hyperparathyroidism was characterised by very low phosphate and a low potassium; patients were all female. Bronchial carcinoma without bone metastases showed less striking phosphate lowering than the hyperparathyroid group and also had very low albumin and raised globulin. The lymphoma patients had normal phosphate values, low globulin and high potassium. They derived discriminant functions based on phosphate, alkaline phosphatase, albumin, globulin and potassium values which assist in classification of the more difficult case where a simple pattern recognition approach proves inadequate. Fraser and her colleagues (1971) have already shown the value of discriminant functions in the diagnosis of hypercalcaemias in younger adults.

Serum parathormone

Serum parathormone levels can now be measured and, if the test is available, it can be of considerable value (Posen *et al*., 1976). Except in parathyroidism where the glands are autonomous, hypercalcaemia should otherwise lead to parathyroid suppression so that serum levels are very low. Thus a normal parathormone level virtually excludes hyperparathyroidism as the cause of hypercalcaemia but is consistent with such causes as bone metastases, lymphoma, myelomatosis, sarcoidosis or hypervitaminosis D. Unfortunately the test is least helpful in distinguishing between the two groups which are biochemically the most similar – hyperparathyroidism and carcinoma without bone metastases. The mechanism of hypercalcaemia in this latter group is thought to be the secretion of parathormone-like peptides by the tumour (usually bronchial carcinoma in elderly patients although many other tumours, particularly hypernephroma, may give rise to the syndrome which Lafferty (1966) has termed pseudo-hyperparathyroidism). This accounts for the considerable biochemical similarities but also means that the parathormone-like material may or may not be detected by the parathormone assay (Posen *et al.*, 1976). The very different characteristics of the biochemistry of the lymphoma patients implies that their hypercalcaemia results from some quite different mechanism.

The steroid suppression test can also help in the recognition of hyperarathyroidism, calcium values failing to fall within 7–10 days of corticosteroid medication whilst they usually do fall in other hypercalcaemias.

We have already referred to the greater diagnostic difficulties when hypercalcaemia is associated with renal failure. In particular, phosphate becomes valueless whereas it has considerable discriminatory value in patients with good renal function. Parathormone measurements can be helpful in this situation. However, as in the case of the diagnosis of bone disease in renal failure, full diagnostic evaluation of hypercalcaemia in renal failure is quite often an academic matter which does not justify full investigation in the elderly patient.

Paget's disease

The possibility of Paget's disease is often suggested by the finding of a normal calcium and phosphate together with an otherwise unexplained elevation of alkaline phosphatase in an elderly patient. This is particularly so if elevation is shown to be due to bone alkaline phosphatase by iso-enzyme studies or shown not to be due to liver disease by the finding of a normal 5′-nucleotidase. Radiology can often readily confirm the diagnosis, the pelvis and spine being the most likely sites for involvement.

However, Paget's disease is asymptomatic in the vast majority of cases and seldom requires treatment. So, although it has a considerable prevalence in elderly patients of approximately 5%, a search for Paget's disease is hard to justify.

CHAPTER SEVEN

Thyroid Function Tests

Thyroid disease has a considerable prevalence of over 5% in elderly patients (Jefferys, 1972). Furthermore, both hypothyroidism and thyrotoxicosis are readily treatable diseases but may otherwise be responsible for considerable morbidity. Indeed, untreated thyrotoxicosis, particularly when cardiac complications occur, is a serious threat to life in elderly patients. Diagnosis is often difficult in old age, however. Hypothyroidism may often be present in a non-specific or atypical way in elderly patients (Bahemuka and Hodkinson, 1975) whilst the appearances of old age itself may resemble those of the disease and so confuse diagnosis (Hill, 1968). Atypical presentations of thyrotoxicosis are well known, for example 'apathetic thyrotoxicosis' (Lancet, 1970), and are especially common in older patients (Rønnov-Jessen and Kirkegaard, 1973).

Thyroid function tests are thus of considerable importance in elderly patients to confirm or refute a clinically suspected diagnosis. However, there is a strong case for the screening of elderly patients for thyroid disease, Bahemuka and Hodkinson (1975) having found 2½% of elderly patients to have thyroid disease which was clinically unsuspected in a screening programme involving two thousand in-patients. Two-thirds of this yield was of hypothyroidism and a third thyrotoxicosis.

The present account of the diagnostic use of the thyroid function tests will therefore pay special attention to the screening aspects of their use in the elderly patient.

The thyroid hormones

Thyroxine (T4) is the main hormone secreted by the thyroid gland and is also the main hormone present in the serum. However, triiodothyronine (T3) is far more active metabolically and is produced peripherally from T4. It has far lower serum concentrations than T4.

Both hormones are very substantially bound to serum proteins. Most of the binding, particularly in the case of T4, is to a specific α-globulin carrier protein, thyroxine-binding-globulin (TBG). Pre-albumin is of lesser importance and mainly binds T3 whilst albumin binds both hormones to a limited extent. The bound hormones are metabolically inactive – it is the small free fractions, amounting to about 0.04% of total T4 and 0.5% of total T3, which determine thyroid status. We have already noted that disturbances of binding proteins are common in elderly patients and this is especially important in the case of thyroid hormone binding, about a quarter of elderly patients admitted to a geriatric department showing disturbances sufficiently gross to affect test interpretation (Jefferys *et al.*, 1972). Correction for binding changes is thus essential if total hormone concentrations are measured.

Protein-bound-iodine

Protein-bound-iodine (PBI) was the first practical estimation of circulating thyroid hormones (T4 + T3) and for many years was the main *in vitro* test of thyroid function. It had the advantages of high precision, low cost and could be automated. Unfortunately, it had poor specificity and was often invalidated by iodine ingestion or contamination. Some 5% of results in Britain were spuriously high for such reasons (McKerron and Howorth, 1971) and so the test has now been largely superseded by the direct measurement of serum T4 which does not share these problems.

Serum T4

This is currently the most widely used test which estimates circulating thryroid hormones. It is a more expensive test than PBI and has poorer precision but, because it is unaffected by iodine contamination, it has proved to be a far more practical test in clinical use. Earlier methods are based on competitive protein binding techniques but, whilst these are the most widely used at present, they may well be replaced by the newer radio-immunoassay methods which offer the prospect of increased precision.

Serum T3

As serum concentrations of T3 are only about a sixtieth of those of T4, its measurement presents considerably greater technical difficulties. However, reliable radio-immunoassay methods which can be applied to unextracted serum are now available in special centres and may soon become more generally available.

T3-uptake tests

In view of the importance of protein binding of the thyroid hormones, T3-uptake tests have been widely used. These tests might more appropriately have been called thyroid hormone binding tests. The protein binding sites of serum are never fully occupied by thyroid hormones even in the thyrotoxic patient. Tests involve the addition of an excess of labelled T3 to serum so as to fully saturate the binding sites. Because T4 is more avidly bound it remains substantially undisturbed. A secondary binding material can then be added (e.g. resin) so as to allow the estimation of either the occupied or the unoccupied binding sites depending on the details of the technique. Thus, in the commonly used Thyopac-3 test (Radiochemical Centre, Amersham), known amounts of a binding resin and radio-iodine labelled T3 are added to the serum specimens and to a standard serum specimen. After equilibration, the radioactivity of the supernatants is counted and expressed as a percentage of that of the standard serum. This T3-uptake result is a measure of unoccupied binding sites. Results are thus typically high in hypothyroidism and low in thyrotoxicosis.

However, T3-uptake results are affected by variations in the thyroid hormone binding proteins, particularly TBG, as well as by changes in thyroid status. T3-uptake levels fall in severe illness (Jefferys, 1972) or with administration of anabolic steroids, or androgens. They rise with oestrogen therapy, in liver disease or after treatment with clofibrate or perphenazine. T4 and T3 levels will be correspondingly high or low in these cases. These misleading variations in total hormone levels can be circumvented by the use of the *free-thyroxine-index* (FTI) or the corresponding free-T3-index in which the hormone value is divided by the T3-uptake result to give a measurement which correlates closely with measurements of the corresponding free hormone concentrations, these last being technically possible but only available as technically difficult research procedures.

The *effective thyroxine ratio* (ETR) has been developed as an alternative to the FTI. This is a combined test which measures T4 with allowance for the protein binding capacity. However, there are some doubts as to whether an equally satisfactory correction is obtained (D'Haene *et al.*, 1974).

Serum thyroid stimulating hormone (TSH)

Serum TSH can be measured by immunoassay and is now fairly generally available as a special if not as a routine investigation. It is not of value in thyrotoxicosis as the sensitivity of the current methods of assay does not allow a useful distinction to be made between normal and low values. In hypothyroidism other than that secondary to hypopituitarism, TSH levels are raised and are of considerable value in confirming the diagnosis. However, a raised TSH value does not automatically mean that the patient is hypothyroid. Thus for example, raised TSH levels are compatible with continuing euthyroid status in patients who have had radioiodine therapy for thyrotoxicosis, although there is a higher associated risk of the subsequent development of hypothyroidism (Toft *et al.*, 1974).

Thyroid function tests in the elderly

Studies of the thyroid function tests in the elderly have been somewhat limited up to the present time. In particular, there have been very few attempts to determine normal ranges appropriate to old age.

Certainly, when thyroid function tests are applied to ill elderly patients, the effects of illness itself as well as those of age need to be considered.

T3-uptake

There is no evidence of any appreciable change in T3-uptake with age, the means for T3-uptake found in the healthy elderly being similar to those in younger subjects (Taylor *et al.*, 1974; Kind and Ghosh, 1976; Hodkinson and Denham, 1977). However, ranges are somewhat wider in older subjects and women have significantly higher values than men although the difference is not of any great clinical importance. Thus for example Hodkinson and Denham (1977) found a 95% range for Thyopac-3 T3-uptake of 93–122% for men and 95–125% for women in healthy euthyroid elderly subjects as compared with the usual 'adult' range of 92–117% quoted by the manufacturers (Radiochemical Centre, Amersham).

In elderly patients, however, low T3-uptake values are often found and are associated with severity of illness (Jefferys *et al.*, 1972). Ranges for T3-uptake in elderly patients are thus lower and wider than for well old people, Jefferys and his colleagues (1972) finding them to be 82–116% for men and 86–119% for women. The fall of T3-uptake with severe illness has been shown in younger patients (Harvey, 1971). Jefferys and his colleagues (1972) showed that the fall in T3-uptake in their elderly patients was accompanied by hypoalbuminaemia and presumably represented a general impairment of protein synthesis as a consequence of illness. As thyroid tests were distorted by such changes in a quarter of elderly patients, the use of the FTI which allows for protein changes is essential. They showed that interpretation was very greatly simplified by its use.

Serum T4

Serum T4 values appear to be little changed in the healthy elderly, Hodkinson and Denham (1977) finding values of 93 ± 18 nmol/litre (7.2 ± 1.4 μg/100 ml). However, there have been recent reports of elevated T4 levels in ill elderly patients, particularly females, who are euthyroid (Britton *et al.*, 1975a; Burrows *et al.*, 1975). Our experience at Northwick Park Hospital suggests that such high T4 values may often be due to drug side effects, with L-Dopa, digitalis glycosides and cotrimoxazole as the main offenders (Baruch *et al.*, 1976). If patients on such drugs are excluded, the range for T4 is closely similar in euthyroid in-patients and the healthy elderly. However, the range for in-patients is rather wider, 93.5 ± 21.3 nmol/litre (7.2 ± 1.6 μg/100 ml), because of the greater variability of binding proteins (Northwick Park Hospital, N = 71).

Serum T3

The serum T3 has yet to be adequately studied in old age. However, lower values than can simply be attributed to decreased TBG levels are commonly found in ill euthyroid elderly patients (Burrows *et al.*,

1975; Britton *et al.*, 1975a). This is probably due to altered peripheral conversion of T4, metabolically inactive reverse-T3 being produced instead of T3 itself (Burger *et al.*, 1976). However, Bray and his colleagues (1976) have shown that T3 levels are related to body weight. Weight loss in elderly patients might thus be an additional factor accounting for low T3 values.

Free-thyroxine-index

Jefferys and his colleagues determined a range for FTI derived from PBI and T3-uptake in elderly patients who were euthyroid (Jefferys *et al.*, 1972). Hodkinson and Denham (1977) found the 95% range for FTI (Thyopac-4 T4 nmol/l ÷ Thyopac-3 T3-uptake %) in healthy euthyroid elderly subjects to be 0.58–1.24. This is a narrower range than that for elderly in-patients using the same tests, 0.54–1.59 (Hodkinson, 1975), the discrepancy probably being mainly due to drug effects as already mentioned when serum T4 was considered.

Serum TSH

The normal range for TSH in the elderly is yet to be determined. Our own experience at Northwick Park Hospital suggests that there may be a significant number of euthyroid elderly people who have moderately elevated TSH levels. Nonetheless, TSH values of 20 μU/litre or more appear to be virtually diagnostic of hypothyroidism in the elderly patient who has not had thyroidectomy or radio-iodine therapy.

Cholesterol

Cholesterol is no longer regarded as a useful screening test for thyroid disease. This view is well founded in elderly patients as Denham (1972a) has shown. He found that elderly patients have an excess of low as well as of high cholesterol values, his 95% ranges being 2.0–8.8 mmol/litre (77–341 mg/100 ml) for men and 2.8–9.2 mmol/litre (108–356 mg/100 ml) for women as compared with ranges of 4.0–7.6 (155–294 mg/100 ml) and 3.7–8.7 mmol/litre (143–337 mg/100 ml) for 122 healthy elderly men and women resident in Harrow. The cholesterol values of his elderly patients were correlated with serum albumin values so that protein binding changes in illness might account for the far greater variability of cholesterol results.

A raised mean-corpuscular-volume (MCV) appears to be a more

helpful indicator of possible hypothyroidism than a high cholesterol in the elderly patient (Bahemuka and Hodkinson, 1975).

Thyroid antibody tests are also of very little practical value in the diagnosis of hypothyroidism or thyrotoxicosis in the elderly patient.

Laboratory diagnosis of thyroid disease in the elderly patient

Despite the many technical advances and great clinical interest in thyroid function tests in recent years, there has been a surprising lack of attention to effective interpretation of the tests, particularly when used in a screening situation. Britton and his colleagues (Britton *et al.*, 1975b) have determined 'decision-aiding ranges' for tests but otherwise there has been a general neglect of probability based methods.

Diagnosis of hypothyroidism using the FTI

Screening the elderly for hypothyroidism is a practical and useful activity (Bahemuka and Hodkinson, 1975). However, the prevalence of clinically unrecognised hypothyroidism is low – around 2% – so that a simple normal range approach to interpretation of FTI would be quite inadequate. Fortunately we can make use of the probabilistic method described in Chapter 2.

The FTI results of euthyroid and hypothyroid elderly subjects are adequately described by normal distributions whose means and standard deviations are 0.86 ± 0.17 (Hodkinson and Denham, 1977) and 0.33 ± 0.19 (Northwick Park Hospital, $N = 78$) respectively. Figure 26 shows the posterior probabilities which can be calculated from the relative probability densities given a prior probability of 0.5 (1 : 1). These posterior probabilities are also given in the form of odds so that they can be adjusted to fit other prior probabilities by simple multiplication as we saw in Chapter 6.

Fig. 27 shows the discrimination achieved when the prior probability is 0.02 – the approximate average prevalence in the elderly patient for clinically inapparent hypothyroidism. The range of values which are equivocal is wide. However, duplicate determinations of FTI give some help, within individual variance for FTI being 0.029 The distributions for duplicate FTI are thus 0.86 ± 0.12 and 0.33 ± 0.14 and Fig. 28 shows the posterior probabilities given by these parameters.

Whilst duplicate FTI determinations are of some value, further

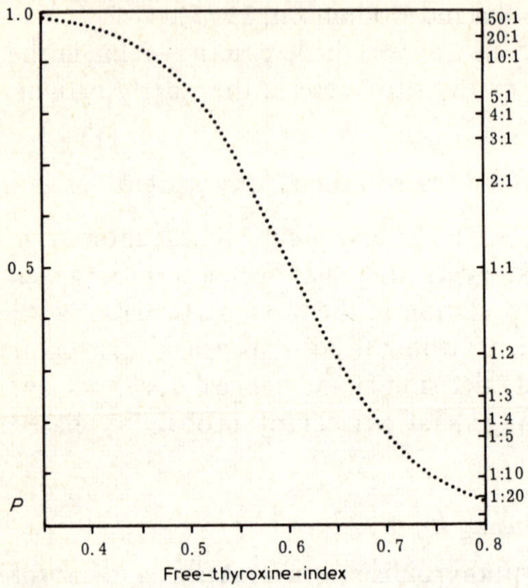

Fig. 26. Posterior probability for myxoedema given by a single FTI result in an elderly patient when prior probability is 0.5.

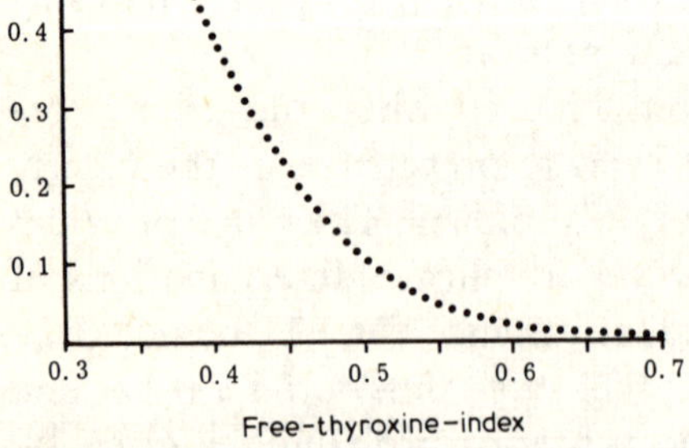

Fig. 27. Posterior probability of myxoedema from a single FTI result when prior probability is 0.02.

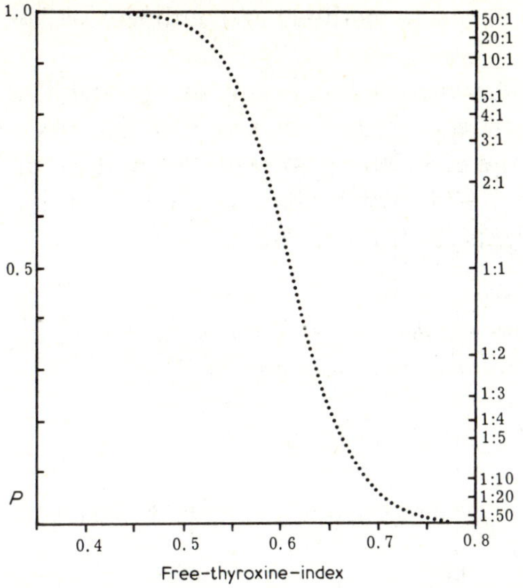

Fig. 28. Posterior probability of myxoedema given by duplicate FTI results when prior probability is 0.5.

repeats are far less useful as we showed in Chapter 2. In an uncertain case it is better to go on to determination of serum TSH which, although not a strictly independent test, has considerably more discriminatory power. Serum T3 measurement is of very dubious value as a further test in the diagnosis of hypothyroidism in the elderly patient.

It should be stressed that the above probability model for the interpretation of FTI results in elderly patients relates to results derived from Thyopac-3 and 4 measurements of T3-uptake and T4. Unfortunately, the considerable variety of techniques which are currently used to measure T3-uptake and T4 give results which are not directly comparable. This is evident when direct comparisons are made (D'Haene *et al*., 1974) but is also shown by the marked differences in normal ranges reported from different centres.

Therefore, whilst the probability model described can be used with reasonable confidence if FTI is derived by the same methods, it needs to be applied with great caution if laboratory methods are different.

Ideally, each centre should determine its own ranges for elderly euthyroid and hypothyroid subjects and calculate the appropriate probabilities. The required statistical methods are outlined in Appendix I.

The other practical problem involved in using the probability method is that of choosing an appropriate prior probability. As a general guide, Table VI gives the prevalences estimated from the data of Bahemuka and Hodkinson (1975) which are based on the routine screening of 2000 elderly patients for thyroid disease.

TABLE VI

Estimated prevalence rates for hypothyroidism in elderly patients without clinically apparent disease (from Bahemuka and Hodkinson, 1975).

Clinical findings	Prevalence of hypothyroidism	
	Females	Males
No relevant associated disease	0.017	0.006
Pernicious anaemia	0.06	0.012
Rheumatoid arthritis	0.07	0.016

Diagnosis of thyrotoxicosis using the FTI

Thyrotoxicosis has a considerably lower prevalence than that of hypothyroidism in the elderly – 1.1% in a series of 2000 patients (Bahemuka and Hodkinson, 1975). A classical clinical picture occurs in only 14% of cases but half of the atypical cases have cardiovascular complications such as congestive cardiac failure, left ventricular failure or atrial fibrillation (Hodkinson, 1975).

Unfortunately, no series of elderly thyrotoxic patients has been reported which is sufficiently large to give a reliable estimate of the range of FTI in the disease and allow a probability model to be derived. However, if 1.50 is taken as the action limit for FTI, it is estimated that approximately 95% of thyrotoxic patients would give results above the limit but that only about a fifth of high FTIs would be due to thyrotoxicosis (Hodkinson, 1975). Spurious elevation due to drugs, for example L-dopa, needs to be excluded. Otherwise further alternative diagnostic tests are needed, serum T3 being the most useful choice (Shalet *et al*., 1975; Britton *et al*., 1975b). Adult normal ranges for serum T3 are approximately 1.0–3.0 nmol/litre (65–195 mg/100 ml).

TSH is of limited value as current assays cannot reliably discriminate betwwen the very low levels in thyrotoxicosis and the lower end of the normal range. The radioactive iodine or technetium uptake tests have been largely superseded by serum T3 measurements.

CHAPTER EIGHT

Blood Sugar and Diabetes

The diagnosis of diabetes relies principally on three tests, the detection of glycosuria, measurement of fasting or random blood sugar and the glucose tolerance test.

Unfortunately, glycosuria is far less useful as a screening test for diabetes in older individuals. Denham (1972b) has shown that almost half of elderly patients with significant hyperglycaemia do not exhibit glycosuria. This appears to be due to the frequent elevation of the renal threshold for glucose in old age. The finding of glycosuria in an elderly subject therefore means that diabetes is very likely. The absence of glycosuria is no guarantee that diabetes is not present, however. Testing the urine for ketones remains an important test in elderly patients.

Changes in glucose metabolism with age

Community studies have shown that average fasting blood sugar levels rise with age by approximately 0.1 mmol/litre (2 mg/100 ml) per decade. Post-prandial values rise rather more by something like 0.2 mmol/litre (4 mg/100 ml) per decade. The most striking changes are in blood sugar values following a glucose load, these showing rises of 0.4–0.7 mmol/litre (8–12 mg/100 ml) per decade (O'Sullivan *et al.*, 1971). High rates of abnormalities are found if conventional standards of normality for blood sugar or the glucose tolerance test are applied to results of elderly populations. These 'unacceptably' high prevalence rates have led to suggestions that diagnosis according to such standards is unrealistic in old age (Malins, 1968).

O'Sullivan (1974) reviews the evidence for the view that the changes in glucose values with age are pathological rather than physiological. The rise in glucose values, though affecting the whole distribution, is more marked at higher values and variance rises more strikingly than the average. Higher blood sugar values are associated with a poorer prognosis in longitudinal community studies (Framingham Study, 1968) and in elderly female patients (Hodkinson, 1975), the effects being shown across the whole range of values, not just at the highest levels. Furthermore, high blood sugar levels in older subjects can be shown to relate to impaired insulin response (Fraser, 1972).

The rise in blood sugar with age thus appears to be mainly a pathological manifestation rather than a physiological one. There is no convincing case for adjusting standards of normality for age; higher blood sugars and impaired glucose tolerance tests are not benign age changes but appear to represent genuine impairment of carbohydrate tolerance.

Random blood sugar as a screening test for diabetes in the elderly

Denham (1972b) has shown that the random (mid-morning) blood sugar is an effective screening test for diabetes in elderly patients. He found the blood sugar values to be bimodally distributed with two overlapping log-normal populations representing normal and diabetic subjects (Fig. 29). He estimated the total prevalence of diabetes to be 11% of which 3% was already known. Assuming the same prior

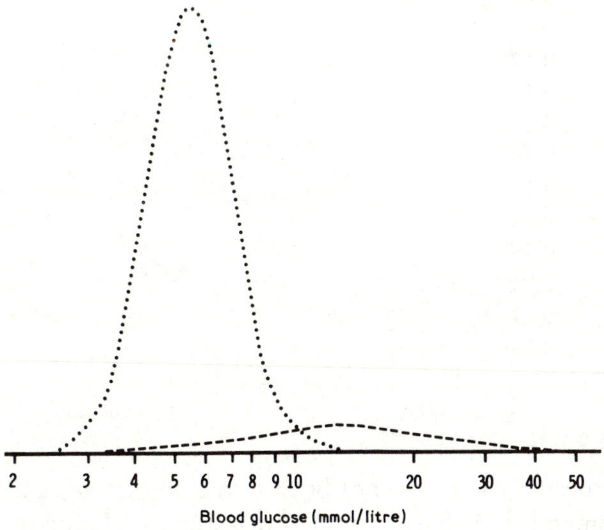

Fig. 29. Blood sugar results in non-diabetic (larger distribution of lower values) and diabetic elderly patients: Logarithmic scale (after Denham, 1972).

probability of 8% for undiagnosed diabetes as he found, Table VII shows the posterior probabilites for diabetes which are implied by different random blood sugar results in elderly patients (mid-morning specimens, venous blood).

If screening were based on the critical limit of 10.7 mmol/litre (193 mg%), 59% of diabetic subjects would be detected whilst only ½% of non-diabetic patients would be wrongly classified. More complete detection would be achieved if a lower action limit were chosen. Thus for example a limit of 9.0 mmol/litre (162 mg/100 ml) would allow 71% of diabetic patients to be recognised at a cost of misclassifying 3% of the non-diabetic patients.

TABLE VII

The probability of diabetes in elderly patients for given random blood sugar values (venous blood)

Mid-morning random blood sugar (mmol/litre)	Probability of diabetes
5.0	0.01
6.0	0.02
7.0	0.03
8.0	0.07
9.0	0.17
9.5	0.24
10.0	0.34
10.5	0.46
10.7	0.50
11.0	0.58
12.0	0.79
13.0	0.91
14.0	0.97
15.0	0.99

The glucose tolerance test

The standards for interpretation have not been shown to differ in old age. It should be remembered that temporary disturbance of glucose tolerance may follow stroke or myocardial infarction. A reasonable delay should be allowed before performing a glucose tolerance test in such circumstances.

Drugs may also provoke the development of impaired glucose tolerance in those who have latent diabetes. The thiazide diuretics, perphenazine, oestrogens and the corticosteriods are the most important examples.

Standards quoted by different authors vary somewhat. Typical upper limits for venous blood glucose values after a 50 gram glucose load are 6.1 mmol/litre (110 mg%) fasting, 9.4 mmol/litre (170 mg%) at 1 hour and 6.7 mmol/litre (120 mg%) at 2 hours (O'Sullivan, 1974). The post-glucose values are somewhat higher if capillary blood is used, for example 10.0 mmol/litre (180 mg%) at 1 hour and 7.5 mmol/litre (135 mg%) at 2 hours (Diabetes Working Party, 1976). Two hour values appear to give better separation of diabetic and non-diabetic subjects than either the fasting or 1 hour values.

Patients whose blood sugar exceeds the 1 hour limit but is below the fasting and 2 hour limits are classified as having lag storage curves and this abnormality is more readily apparent if capillary blood specimens are taken. Lag storage is more common in elderly people and has been regarded as a benign abnormality. However, follow up shows that the risk of mortality and of the subsequent development of diabetes are higher in such individuals (Diabetes Working Party, 1976).

Hypoglycaemia

Hypoglycaemia is an important cause of confusion or disturbance of consciousness in the elderly patient. The lower limit for fasting blood sugar is usually quoted as being approximately 2.5 mmol/litre (45 mg/100 ml) using specific analytical methods such as the glucose oxidase technique. Denham (1972b) found a lower limit of 3.4 mmol/litre (61 mg/100 ml) for random blood sugar in elderly patients.

Most often, hypoglycaemia is due to insulin or oral hypoglycaemic drugs given for diabetes but it may occasionally be seen in liver disease or be due to an insulinoma in the elderly patient. Hypopituitarism and Addison's disease are very rarely responsible in old age.

CHAPTER NINE

Liver Function and Enzyme Tests

There are a great number of liver function tests and enzyme measurements available to the clinician. Many, however, are of questionable usefulness and there is much to be said for a policy of test reduction, i.e. choosing a minimum of tests which have the maximum discriminatory power (Card, 1975). Knill-Jones (1975) found that two-thirds of tests employed in the investigation of liver disease were redundant, only four biochemical tests being necessary for effective discrimination.

The present chapter will thus deal mainly with the tests which are commonly included in biochemical profiles – serum bilirubin, serum glutamic-oxaloacetic transaminase (SGOT) and lactate dehydrogenase (LDH). Other tests will be dealt with only very briefly.

Serum bilirubin

The normal range for serum bilirubin is generally given as 1–15 μmol/litre (0.05–0.9 mg/100 ml) and the findings of Reed and his colleagues (1972) and Sharland (1972) suggest that it is no different in old age. Experience at Northwick Park Hospital supports this view rather than that of Leask and his colleagues (1973) who reported somewhat higher levels in the elderly. There is general agreement that values are a little higher in males at all ages and that results are log-normally distributed. Serum bilirubin is almost entirely bound to albumin and so might be expected to be somewhat lower in the hypoalbuminaemic ill elderly patient. However, such effects and the difference between the sexes appear to be of little practical importance in the interpretation of bilirubin results in the elderly patient.

Spurious elevation of the serum bilirubin as a result of *in vitro* haemolysis may sometimes give rise to confusion, however. Whilst this may be attributable to faulty technique in the collection or storage of the blood specimen, it also occurs for reasons that are poorly understood in some seriously ill elderly patients (M. J. Denham, personal communication).

The further biochemical investigation of the elderly patient with hyperbilirubinaemia differs little from that of the yonger jaundiced patient, at least in principle. However, enzyme tests and serum protein changes are far from specific to liver disease. Misleading alterations may well arise when, as is so often the case, multiple diseases are present in the elderly patient. This problem has already been discussed in respect of the use of alkaline phosphatase as an aid to the diagnosis of osteomalacia (Chapter 6) but equally applies when the test is used in the further investigation of the jaundiced patient. Similarly, hepatocellular disease is only one of many possible causes of hypoalbuminaemia or elevation of serum γ-globulin (Chapter 4). The same problems apply to interpretation of SGOT as we shall see below.

The differential diagnosis of jaundice is also somewhat altered in old age. Obstructive causes predominate. Whilst obstructive jaundice is most often due to stone, malignant causes are very common and include carcinoma of the head of the pancreas, carcinoma of stomach with glands in the porta hepatis, massive hepatic secondaries from a variety of primary sites or, more rarely, cancer of the ampulla, bile duct or gall-bladder. Cholestatic jaundice due to chlorpromazine is less often seen nowadays as the unsuitability of the drug for use in elderly patients is more widely appreciated.

Hepatocellular jaundice is far less common, alcoholic cirrhosis being the form most often seen in the elderly. Viral hepatitis is seldom seen.

Haemolytic anaemias are a rare cause of jaundice in old age. However, mild jaundice due to haemolysis is quite common in pernicious anaemia. It may also occur when pulmonary emboli occur in patients with cardiac failure and with septicaemias. As in other age groups, jaundice can be recognised as being due to haemolysis by the finding of an excess of urobilinogen and the absence of bilirubin on testing the urine.

Further biochemical investigation is therefore principally concerned with the distinguishing of obstructive from hepatocellular jaundice, though both may coexist. Elevation of γ-globulin is not particularly helpful in the older patient as it occurs both in hepatocellular jaundice and obstruction due to malignant disease as well as in many non-hepatic conditions. Flocculation tests are not generally useful; measurement of serum albumin and globulin and electrophoresis of serum proteins where indicated are to be preferred.

Enzyme tests are of greater value, alkaline phosphatase and SGOT being the most useful. Elevation of alkaline phosphatase is typical of obstructive jaundice. However, as we saw in Chapter 6, it is often elevated because of bone abnormalities in elderly patients. Isoenzyme studies can help to distinguish bone from liver origin of a raised alkaline phosphatase. Alternatively the 5′-nucleotidase may be determined. This tends to be high in obstructive biliary disease but is unaffected by bone pathologies.

SGOT

SGOT (also called aspartate transaminase) is raised, often to very high levels, in hepatocellular disease but more modest degrees of elevation may also occur in obstructive jaundice.

Normal range for SGOT appears to be unchanged in older age groups (Reed *et al.*, 1972). However, though the test remains as sensitive an indicator of hepatocellular damage as in other age groups, its poor specificity makes its interpretation far more difficult in the elderly patient. Indeed, 18% of all patients admitted to a geriatric department had SGOT values above the upper limit of 35 i.u. (Northwick Park Hospital, $N = 315$). Whilst a good proportion of the high results could be attributed to liver or cardiac pathology, many were of uncertain significance. SGOT is therefore of rather dubious value as a profile test in elderly patients. Some of the unexplained high results could be due to minor degrees of haemolysis of blood specimens, for the concentration of SGOT and other transaminases is very high in red blood cells.

Serum glutamic-pyruvic transaminase (SGPT), though tending to show greater degrees of elevation in hepatocellular disease and lesser disturbance in myocardial infarction, suffers from comparable problems and offers no clear advantage over SGOT.

'Cardiac' enzymes

SGOT, lactate dehydrogenase (LDH), α-hydroxy-butyric dehydrogenase (HBD) and creatine phosphokinase (CPK) are much used in the diagnosis of myocardial infarction, all the enzymes being released from damaged myocardial cells. All have specificity problems and these are reduced rather than solved when isoenzymes are utilized. SGOT is, of course, elevated in liver disease. Indeed, hepatic congestion in congestive heart failure occurs often enough to give some elevation of SGOT. Alternatively, elevation of SGOT may occur where there is damage of muscle, in pulmonary infarction or following stroke (Denham and Hodkinson, 1973). LDH and HBD are closely related, the latter largely representing the cardiac isoenzyme (isoenzyme I) of LDH and so having improved specificity. Nonetheless HBD, like LDH, may be elevated in liver disease, megaloblastic anaemia, muscle cell injury, pulmonary infarction or stroke as well as in myocardial infarction. Haemolysis, as with SGOT and SGPT, also leads to spuriously elevated values.

CPK is unaffected by liver dysfunction but may be elevated even more readily by muscle damage, intramuscular injection being enough to give elevation (Scott *et al*., 1974). It is the enzyme most often found to be elevated after stroke (Denham and Hodkinson, 1973). Elevation may also occur with pulmonary embolism or pulmonary oedema.

Special caution is necessary in the interpretation of 'cardiac' enzyme results in old people. Normal ranges appear to be no different than in younger age groups (Reed *et al*., 1972). However, the intrinsic lack of specificity of the enzymes is especially important in this age group because multiple disease is so commonly present. Elevation of 'cardiac' enzymes should never be regarded as more than suggestive of myocardial infarction. Confirmatory ECG changes must be found before a firm diagnosis of myocardial infarction can be made. Serial enzyme measurements may help to gain the maximum amount of diagnostic information from these tests, however (Scott *et al*., 1974).

APPENDIX I

Statistical Methods

Inter-laboratory variation means that test ranges in health or disease which have been determined in other centres need to be used only with great caution. It is far better if ranges can be determined locally in a series of appropriate subjects of suitable age and sex.

The statistical methods which are necessary to find ranges and to calculate critical limits or posterior probabilities from them are not particularly difficult. All the necessary mathematical work is well within the capability of modern 'scientific' pocket calculators.

Determination of ranges

Ranges can be estimated from samples of any size. However, the reliability of the estimates of the mean and standard deviation are proportional to $1/\sqrt{N}$ and $1/\sqrt{(2N)}$ respectively, where N is the size of the sample. Reasonably large samples are thus to be preferred but a local range found from a relatively small sample may well be better than one based on larger samples elsewhere.

It is important that specimen collection and analysis keep as closely as possible to the usual clinical practice. Technique and time of venepuncture should be comparable and analysis should be spread over many routine batches, not grouped into one or a few special batches. Otherwise analytical variation will be misleadingly lower than it is in everyday use of the test, and the range unrealistically restricted.

Calculation of the mean is very simple whilst determination of the standard deviation presents no great difficulties. The formula:

$$\text{standard deviation } (s) = \sqrt{\left(\frac{\Sigma(\bar{x} - x)^2}{(N-1)}\right)}$$

has already been given in Chapter 2. However, standard deviation may be rather more conveniently calculated on a calculator from an alternative form of the formula:

$$s = \sqrt{\left(\frac{\Sigma x^2 - (\Sigma x)^2/N}{(N-1)}\right)}$$

This has the advantage of not requiring prior calculation of the mean and so needs only a single entry of each value to calculate both mean and standard deviation if a calculator with at least two memories is available so that Σx and Σx^2 can be accumulated in them.

Testing for normality

Mean and standard deviation can only be used to estimate normal range or in other statistical applications if the sample values are normally distributed (or nearly so). The most useful test for normality is cumulative probability plotting using probability paper. This is special graph paper in which one scale is a percentage probability scale, i.e. is proportional to the cumulative percentage area under the normal curve. This scaling has the effect of converting what would otherwise be a sigmoid curve when observations from a normal population are plotted cumulatively to a straight line.

The procedure is to first rank the observations in order of increasing magnitude and then to successively plot test value against the cumulative percentage given by the formula:

$$\frac{\text{rank number} - \frac{1}{2}}{\text{total number of observations}}$$

A sample drawn from a normal population should give a reasonably close approximation to a straight line as in the example shown in Fig. 30. If, instead of forming a straight line, the points form a curve which is convex upwards a log-normal distribution is suggested. In this case a change to a logarithmic scale will give a straight line. Standard statistical textbooks may be consulted for details of more searching tests for normality such as χ^2-goodness-of-fit and testing for skewness and kurtosis using the third and fourth moments about the mean. These tests will not usually be necessary, however.

Plotting may also show one or more results which are widely separated from the main population. These outlying results should be deleted and mean and standard deviation recalculated from the remaining results if it can be shown that they are due to laboratory

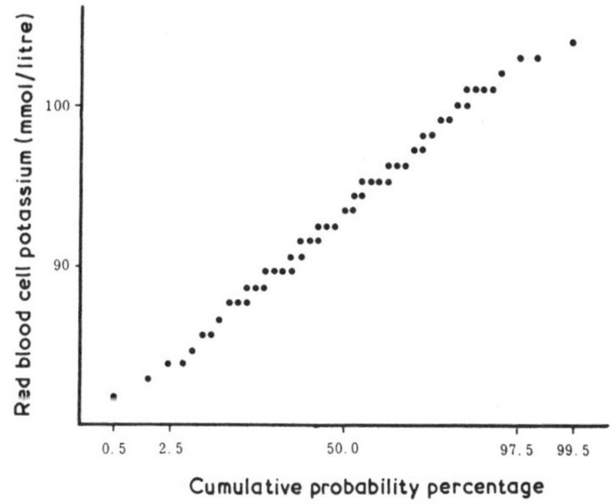

Fig. 30. Cumulative probability plot of red-blood-cell potassium results showing a good fit with the normal distribution (from Bahemuka and Hodkinson, 1975b).

or other errors or to misclassification of the individual as being healthy or having disease. Otherwise an outlying result should only be deleted if t-testing shows the result to have a very low probability of having come from the main distribution. To do this the statistic $(x - \bar{x})/s$ is referred to tables of Student's t with $(N - 1)$ degrees of freedom.

Calculation of posterior probabilities

We saw how a knowledge of the distribution of test results in health and disease may be used to determine posterior probabilities of disease in Chapter 2. These derive from relative probability densities of the two normal distributions at a given test value, i.e.

$$\text{posterior probability of disease} = \frac{\text{probability density disease}}{\text{combined probability densities}}$$

The probability density at a given value x for a normal distribution with mean $\bar{x}$ and standard deviation s is:

$$\text{probability density} = \frac{1}{s\sqrt{(2\pi)}}\, e^{-1/2\left(\frac{\bar{x} - x}{s}\right)^2}$$

When we take the ratio of probability densities to obtain a posterior probability, we need to scale each result in proportion to the appropriate prior probability for disease or non-disease. The term

$1/\sqrt{(2\pi)}$ occurs in both the numerator and denominator and so can be omitted. The formula for the posterior probability for disease given a prior probability of p is thus:

$$\frac{\dfrac{p}{s_1} e^{-1/2\left(\dfrac{\bar{x}_1 - x}{s_1}\right)^2}}{\left[\dfrac{p}{s_1} e^{-1/2\left(\dfrac{\bar{x}_1 - x}{s_1}\right)^2}\right] + \left[\dfrac{(1-p)}{s_2} e^{-1/2\left(\dfrac{\bar{x}_2 - x}{s_2}\right)^2}\right]}$$

It should be noted that the left hand term in the denominator is identical to the numerator of the formula and need only be calculated once.

Often the two standard deviations s_1 and s_2 are equal or nearly so and this is invariably so in the case of discriminant functions. The formula becomes somewhat simpler as the terms p/s_1 and $(1-p)/s_2$ reduce to p and $(1-p)$. Although superficially somewhat daunting, the formula can be calculated easily with the aid of a 'scientific' pocket calculator. Repeated determinations of posterior probabilities from different values of x and p can be calculated very rapidly and easily if the calculator is one which can be programmed.

APPENDIX II

Unpublished Biochemical Studies from Northwick Park Hospital

The various unpublished studies referred to in the text relate to information gained from the analysis of routine biochemical profile results of patients admitted to the department of geriatric medicine during the period 1970–1976.

The pattern of work of the department has been more fully described elsewhere (Hodkinson and Jefferys, 1972). The salient features are that there is usually no waiting list for admission so that those admitted include many old people who have acute medical problems. Lengths of stay are generally short as is shown in Figs. 31 and 32.

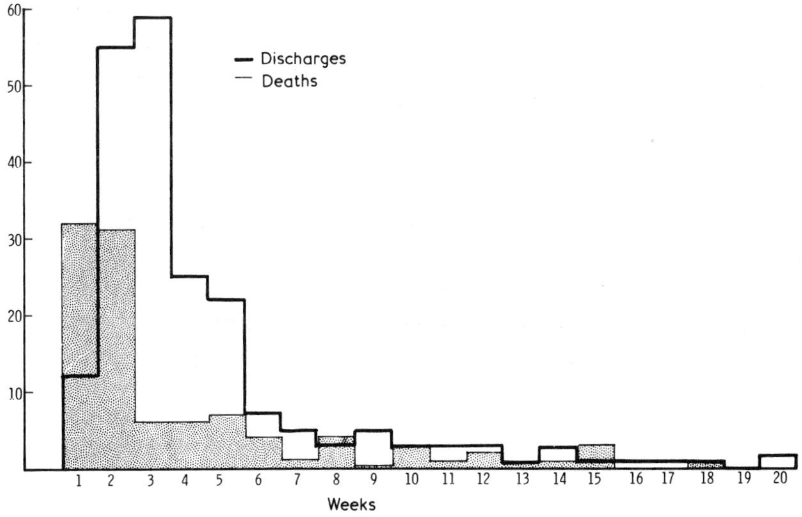

Fig. 31. Pattern of discharge and death in male patients admitted to the department of geriatric medicine, Northwick Park Hospital.

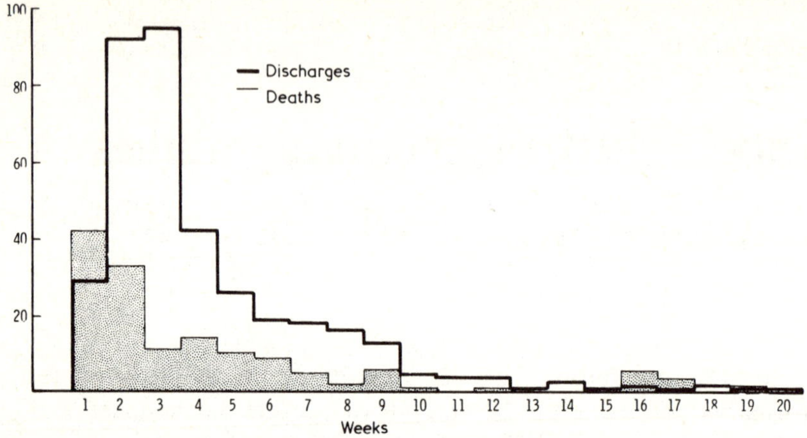

Fig. 32. Pattern of discharge and death in female patients admitted to the department of geriatric medicine, Northwick Park Hospital.

Patients are therefore rather more representative of the elderly admitted to all hospital medical departments than is generally the case for a department of geriatric medicine.

All patients are aged sixty or over, average age being 81 for women and 78 for men. The women admitted outnumber the men by approximately two to one.

Blood for routine profile tests is taken at around 10 a.m. on the first weekday morning following the patient's admission. Venepuncture is performed by experienced technicians, trained to use minimal venous cuffing. Patients are not fasted and have had breakfast some time before venepuncture.

Profile results are obtained from more than 95% of patients admitted, those not investigated largely consisting either of patients who die shortly after admission or have been recently investigated so that further tests are thought unnecessary. The results are therefore free from any major bias due to selection.

Technical details of laboratory methods have been detailed elsewhere (Hodkinson, 1975).

APPENDIX III

SI Units

Fuller details of SI units (Système International d'Unités) are to be found in recent publications by Baron (1974) and Paterson and Frazer (1975).

In the context of the reporting of biochemical results, the main changes from previous practice relate to the expression of concentrations in molar terms and the replacement of the millilitre and 100 millilitres by the litre as the base for volumetric concentrations. The submultiples of the mole relevant to biochemical reporting are the millimole (mmol; $mole^{-3}$), micromole (μmol; $mole^{-6}$), nanomole (nmol; $mole^{-9}$) and picomole (pmol; $mole^{-12}$).

Where molar concentrations are inapplicable or inappropriate, for example for mixtures such as serum globulin, the gram or its

TABLE VIII

Conversions to SI units

Test	Old units	SI units	Conversion factor from old to new units
Serum albumin	g/100 ml	g/litre	x10
Serum globulin	g/100 ml	g/litre	x10
Electrolytes	meq/litre	mmol/litre	identical
Urea	mg/100 ml	mmol/litre	÷6
Creatinine	mg/100 ml	μmol/litre	x88
Uric acid	mg/100 ml	μmol/litre	x60
Calcium	mg/100 ml	mmol/litre	÷4
Calcium	meq/litre	mmol/litre	÷2
Phosphate	mg/100 ml	mmol/litre	÷3.1
T4	μg/100 ml	nmol/litre	x13
T3	ng/100 ml	nmol/litre	÷65
Bilirubin	mg/100 ml	μmol/litre	x17
Glucose	mg/100 ml	mmol/litre	÷18

submultiples are used whilst international units per litre (iu/litre) are still used for enzyme results at the present time.

Some examples of the relationship between older units and SI units are shown in Table VIII.

References

Anderson, I., Campbell, A. E. R., Dunn, A. and Runciman, J. B. M. (1966), Osteomalacia in elderly women. *Scot. Med. J.*, **2**, 555.

Bahemuka, M., Hodkinson, H. M., Denham, M. J. and Padmore, G. R. A. (1974), Serum creatinine in a geriatric inpatient population. *Age & Ageing*, **3**, 43.

Bahemuka, M. and Hodkinson, H. M. (1975), Screening for hypothyroidism in the elderly patient. *Brit. Med. J.*, **2**, 601.

Bahemuka, M. and Hodkinson, H. M. (1976a), Red-blood-cell potassium as a practical index of potassium status in elderly patients. *Age & Ageing*, **5**, 24.

Bahemuka, M. and Hodkinson, H. M. (1976b), Red-blood-cell potassium and hand-grip strength in healthy elderly people. *Age & Ageing*, **5**, 116.

Barnett, R. N., Civin, W. H. and Schoen, I. (1970), Multiphasic screening by laboratory tests – an overview of the problem. *Amer. J. Clin. Path.*, **54**, 483.

Baron, D. N. (1974), SI units. *Brit. Med. J.*, **1**, 509.

Bartter, F. C. and Schwartz, W. B. (1967), Inappropriate secretion of antidiuretic hormone. *Amer. J. Med.*, **42**, 790.

Baruch, A. L. H., Davis, C. and Hodkinson, H. M. (1976), Causes of high free-thyroxine-index values in sick euthyroid elderly patients. *Age & Ageing*, **5**, 224.

Bates, B. and Yellin, J. A. (1972), The yield of multiphasic screening. *J. Amer. Med. Assn.*, **222**, 74.

Betro, M. G. and Pain, R. W. (1972), Hypophosphataemia and hyperphosphataemia in a hospital population. *Brit. Med. J.*, **1**, 273.

Boyd, R. V. and Hoffbrand, B. I. (1966), Erythrocyte-sedimentation-rate in elderly hospital patients. *Brit. Med. J.*, **1**, 901.

Bradwell, A. R., Carmalt, M. H. B. and Whitehead, T. P. (1974), Explaining the unexpected abnormal results of biochemical profile investigations. *Lancet*, **2**, 1071.

Bray, G. A., Fisher, D. A. and Chopra, I. J. (1976), Relation of thyroid hormones to body weight. *Lancet*, **1**, 1206.

Britton, K. E., Quinn, V., Ellis, S. M., Cayley, A. D. C., Miralles, J. M., Brown, B. L. and Ekins, R. P. (1975a), Is 'T4-toxicosis' a normal biochemical finding in elderly women? *Lancet*, **2**, 141.

Britton, K. E., Quinn, V., Brown, B. L. and Ekins, R. P. (1975b), A strategy for thyroid function tests. *Brit. Med. J.*, **3**, 350.

Brown, I. R. F., Bakowska, A. and Millard, P. H. (1976), Vitamin D status of patients with femoral neck fracture. *Age & Ageing*, **5**, 127.

Burger, A., Nicod, P., Suter, P., Vallotton, M. B., Vagenakis, A. and Braverman, A. (1976), Reduced active thyroid hormone levels in acute illness. *Lancet*, **1**, 653.

Burrows, A. W., Shakespear, R. A., Hesch, R. D., Cooper, E., Aickin, C. M. and Burke, C. W. (1975), Thyroid hormones in the elderly sick: 'T4 euthyroidism'. *Brit. Med. J.*, **4**, 437.

Card, W. I. (1975), Choice of diagnostic tests. *J. Roy. Coll. Physicians, Lond.*, **9**, 264.

Carter, P. M., Davison, A. J., Wickins, H. I. and Zilva, J. F. (1974), Quality and quantity in chemical pathology. *Lancet*, **2**, 1555.

Chalmers, J., Conacher, W. D. H., Gardner, D. L. and Scott, P. J. (1967), Osteomalacia – a common disease in elderly women. *J. Bone & Jt. Surg.*, **49B**, 403.

Corless, D., Beer, M., Boucher, B. J., Gupta, S. P. and Cohen, R. D. (1975), Vitamin D status in long-stay geriatric patients. *Lancet*, **1**, 1404.

Davidson, C., Burkinshaw, L., McLachlan, M. S. F. and Morgan, D. B. (1976), Effects of long-term diuretic treatment on body potassium in heart disease. *Lancet*, **2**, 1044.

Denham, M. J. (1972a), Cholesterol as a parameter of subnutrition in the elderly inpatient. *Gerontologia clinica*, **14**, 235.

Denham, M. J. (1972b), The value of random blood glucose determinations as a screening method for detecting diabetes mellitus in the elderly patient. *Age & Ageing*, **1**, 55.

Denham, M. J. (1973), Another cause for confusion. *Modern Geriatrics*, **3**, 332.

Denham, M. J. and Himsworth, R. L. (1974), Hyperthyroidism induced by potassium iodide given in the course of ^{125}I-fibrinogen test. *Age & Ageing*, **3**, 221.

Denham, M. J. and Hodkinson, H. M. (1973), Serum enzymes and prognosis in cerebral infarction. *Age & Ageing*, **2**, 86.

Denham, M. J., Hodkinson, H. M. and Fisher, M. (1975), Glomerular filtration rate in sick elderly inpatients. *Age & Ageing*, **4**, 32.

Dent, C. E., Norris, T. St. M., Smith, R., Sutton, R. A. L. and Temperley, J. M. (1968), Steatorrhoea with striking increase of plasma alkaline phosphatase of intestinal origin. *Lancet*, **1**, 1333.

D'Haene, E. G. M., Crombag, F. J. L. and Tertoolen, J. F. W. (1974), Comparison between determinations of free-thyroxine-index and effective-thyroxine-ratio in human serum. *Brit. Med. J.*, **3**, 708.

Diabetes Working Party (1976), Ten year follow-up report on Birmingham Diabetes Survy of 1961. *Brit. Med. J.*, **2**, 35.

Diem, K. and Lentner, C. (1970) (Eds.), *Scientific Tables*, Documenta Geigy, J. R. Geigy, Basle.

Dixon, W. J. (1967), *BMD biomedical computer programs*, University of California Press, Berkeley & Los Angeles.

Elveback, L. R., Guiller, C. L. and Keating, F. R. (1970), Health, normality and the ghost of Gauss. *J. Amer. Med. Assn.*, **211**, 69.

Flynn, F. V., Piper, K. A. J., Garcia-Webb, P., McPherson, K. and Healy, M. J. R. (1974), Frequency distributions of commonly determined blood constituents in healthy blood donors. *Clin. Chim. Acta*, **52**, 163.

Flynn, F. V., Piper, K. A. J., Garcia-Webb, P., McPherson, K. and Healy, M. J. R. (1976), Biological and analytical variation of commonly determined blood constituents in healthy blood donors. *Clin. Chim. Acta*, **70**, 179.

Framingham Study (1968), *An epidemiological investigation of cardiovascular disease.* U.S. Government Printing Office, Washington D.C.

Fraser, R. (1972), Metabolic disorders in diabetes. *Brit. Med. J.*, **4**, 591.

Fraser, P., Healy, M., Rose, N. and Watson, L. (1971), Discriminant functions in the differential diagnosis of hypercalcaemia. *Lancet*, **1**, 1314.

Grero, P. S. and Hodkinson, H. M. (1977), Hypercalcaemia in elderly hospital inpatients: value of discriminant analysis in differential diagnosis. *Age & Ageing*, **6**, 14.

Hall, G. H. and Hargreaves, T. (1972), Giant-cell arteritis and raised serum alkaline phosphatase levels. *Lancet*, **2**, 48.

Harris, E. K. (1974), Effects of intra- and interindividual variation on the appropriate use of normal ranges. *Clin. Chem.*, **20**, 1535.

Harvey, R. F. (1971), Serum thyroxine and thyroxine-binding-globulin in seriously ill patients. *Lancet*, **1**, 208.

Hazell, K. and Ortiz, S. (1966), The blood alkaline phosphatase level as an aid to diagnosis, treatment and prognosis. *Gerontologia clinica*, **8**, 111.

Hill, W. C. (1968), The aged thyroid gland. *Geriatics*, **23**, 124.

Hobson, W. and Jordan, A. (1959), A study of serum alkaline phosphatase levels in old people living at home. *J. Gerontology*, **14**, 292.

Hodkinson, H. M. (1973), Non-specific presentation of illness. *Brit. Med. J.*, **4**, 94.

Hodkinson, H. M. (1973a), Serum inorganic phosphate in a geriatric inpatient population. *Gerontologia clinica*, **15**, 45.

Hodkinson, H. M. (1973b), Serum calcium in a geriatric inpatient population. *Age & Ageing*, **2**, 157.

Hodkinson, H. M. (1973c), Interpretation of biochemical data in the elderly. *Brit. J. Hosp. Med.*, **10**, 684.

Hodkinson, H. M. (1974), Biochemical side effects of drugs in the elderly. *Gerontologia clinica*, **16**, 175.

Hodkinson, H. M. (1975), Diagnostic and prognostic aspects of routine laboratory screening of the geriatric inpatient. *D.M. Thesis*, University of Oxford.

Hodkinson, H. M. and Jefferys, P. M. (1972), Making hospital geriatrics work. *Brit. Med. J.*, **4**, 536.

Hodkinson, H. M. and Pomerance, A. (1974), Cardiac amyloidosis in the elderly: associated elevation of serum alkaline phosphatase. *Age & Ageing*, **3**, 76.

Hodkinson, H. M. and McPherson, C. K. (1973), Alkaline phosphatase in a geriatric inpatient population. *Age & Ageing*, **2**, 28.

Hodkinson, H. M., Darby, A. and Slavin, G. (1977), Biochemical diagnosis of osteomalacia: a clinicopathological study. In preparation.

Hodkinson, H. M. and Denham, M. J. (1977), Thyroid function tests in the elderly in the community. *Age & Ageing*, **6**, 67.

Holland, W. W. and Whitehead, T. P. (1974), Value of laboratory tests in diagnosis and treatment. *Lancet*, **2**, 391.

Jefferys, P. M. (1972), Prevalence of thyroid disease in patients admitted to a geriatric department. *Age & Ageing*, **1**, 33.

Jefferys, P. M., Farran, H. E. A., Hoffenberg, R., Fraser, P. M. and Hodkinson, H. M. (1972), Thyroid function tests in the elderly. *Lancet*, **1**, 924.

Kampmann, J., Siersback-Nielsen, K., Kristensen, M. and Mølholm-Hansen, J. (1974), Rapid evaluation of creatinine clearance. *Acta Med. Scand.*, **196**, 517.

Keating, F. R., Jones, J. D., Elveback, L. R. & Randall, R. V. (1969), The relation of age and sex to distribution of values in healthy adults of serum calcium, inorganic phosphorus, magnesium, alkaline phosphatase, total proteins, albumin and blood urea. *J. Lab. Clin. Med.*, **73**, 825.

Kind, P. R. N. and Ghosh, S. (1976), Observations on thyroid function tests in the elderly. *Age & Ageing*, **5**, 141.

Knill-Jones, R. P. (1975), The diagnosis of jaundice by the computation of probabilities. *J. Roy. Coll. Physicians, Lond.*, **9**, 205.

Knox, E. G. (1974), Multiphasic screening. *Lancet*, **2**, 1434.

Kodicek, E. (1974), The story of vitamin D from vitamin to hormone. *Lancet*, **1**, 325.

Lafferty, F. W. (1966), Pseudohyperparathyroidism. *Medicine*, **45**, 247.

Lancet (1970), Apathetic thyrotoxicosis. *Lancet*, **2**, 809.

Leask, R. G. S., Andrews, G. R. and Caird, F. I. (1973), Normal values of sixteen blood constituents in the elderly. *Age & Ageing*, **2**, 14.

Lewis, A. E. (1966), *Biostatistics*, Reinhold Publishing Co., New York.

Malins, J. (1968), *Clinical Diabetes Mellitus*, Eyre & Spottiswoode, London.

Maugham, E., Williams, J. R. B., Pomerance, A. and Hodkinson, H. M. (1977), Immunoglobulins in senile cardiac amyloidosis. To be published.

McKerron, C. G. and Howorth, P. J. N. (1971), Thyroid function tests. *Update*, **3**, 1409.

Middler, S., Pak, C. Y. C., Murad, F. and Bartter, F. C. (1973), Thiazide diuretics and calcium metabolism. *Metabolism*, **22**, 139.

Mills, J. N. (1966), Human circadian rhythms. *Physiol. Rev.*, **46**, 128.

Mukherjee, A. P., Coni, N. K. and Davison, W. (1973), Osmoreceptor function among the elderly. *Gerontologia clinica*, **15**, 227.

Murphy, E. A. and Abbey, H. (1967), The normal range – a common misuse. *J. Chronic Dis.*, **20**, 79.

Nordin, B. E. C. (1973), *Metabolic Bone and Stone Disease*, Churchill Livingstone, Edinburgh and London.

O'Sullivan, J. B. (1974), Age gradient in blood sugar levels. *Diabetes*, **23**, 713.

O'Sullivan, J. B., Mahan, C. M., Freedlender, A. E. and Williams, R. F. (1971), The effect of age on carbohydrate metabolism. *J. Clin. Endocrinol. Metab.*, **33**, 619.

Parfitt, A. M. (1965), Changes in serum calcium and phosphate during stilboestrol treatment of osteoporosis. *J. Bone & Jt. Surg.*, **47B**, 137.

Paterson, N. and Frazer, S. C. (1975), The international system of units (SI units) in medicine. *Brit. J. Hosp. Med.*, **12**, 759.

Peterson, P. A. (1971), Isolation and partial characterisation of a human vitamin D-binding plasma protein. *J. Biol. Chem.*, **246**, 7748.

Posen, S., Kleerekoper, M., Ingham, J. P. and Hirshorn, J. E. (1976), Parathyroid hormone assay in clinical decision-making. *Brit. Med. J.*, **1**, 16.

Reece, R. L. (1974), Using a computer to interpret multiphasic screening results. *Geriatrics*, **29**, 51.

Reed, A. H., Cannon, D. C., Winkelman, J. W., Bhasin, Y. P., Henry, R. J. and Pileggi, V. J. (1972), Estimation of normal ranges from a controlled sample survey: I Sex and age related influence on SMA 12/60 screening group of tests. *Clin. Chem.*, **18**, 57.

Riggs, B. L., Jowsey, J., Kelly, P. J., Jones, J. D. and Maher, F. T. (1969), Effects of sex-hormones on bone in primary osteoporosis. *J. Clin. Invest.*, **48**, 1065.

Roberts, L. B. (1967), The normal ranges, with statistical analysis for seventeen blood constituents. *Clin. Chim. Acta*, **16**, 69.

Rønnov-Jessen, V. and Kirkegaard, C. (1973), Hyperthyroidism – a disease of old age? *Brit. Med. J.*, **1**, 41.

Scott, B. B., Simmons, A. V., Newton, K. E. and Payne, R. B. (1974), Interpretation of serum creatine kinase in suspected myocardial infarction. *Brit. Med. J.*, **4**, 691.

Shalet, S. M., Beardwell, C. G., Lamb, A. M. and Gowland, E. (1975), Value of routine serum-triiodothyronine estimation in the diagnosis of thyrotoxicosis. *Lancet*, **2**, 1008.

Sharland, D. E. (1972), Serum alkaline phosphatase: the levels and patterns of iso-enzymes in the non-hospitalised elderly. *Age & Ageing*, **1**, 168.

Sharland, D. E. (1977), Personal communication.

Sourander, L. B. (1973), in *Textbook of Geriatric Medicine and Gerontology* (Ed. Brocklehurst, J. C.), Churchill Livingstone, Edinburgh and London.

Stoker, D. J., Wynn, V. and Robertson, G. (1966), Effect of posture on plasma cortisol level. *Brit. Med. J.*, **1**, 336.

Taylor, B. B., Thomson, J. A. and Caird, F. I. (1974), Further studies of thyroid function tests in the elderly at home. *Age & Ageing*, **3**, 122.

Thomas, J. H. and Powell, D. E. B. (1971), *Blood Disorders in the Elderly*, J. Wright & Sons Ltd., Bristol.

Toft, A. D., Irvine, W. J., Hunter, W. M. and Seth, J. (1974), Plasma TSH and serum T4 levels in long-term follow-up of patients treated with ^{131}I for thyrotoxicosis. *Brit. Med. J.*, **3**, 152.

Whitehead, T. P. and Wootton, I. D. P. (1974), Biochemical profiles for hospital patients. *Lancet*, **2**, 1439.

Woodford-Williams, E., Webster, D. and Landless, B. (1966), Serum protein studies in myocardial infarction. *Gerontologia clinica*, **8**, 44.

Young, D. S., Thomas, D. W., Friedman, R. B. and pestaner, L. C. (1972), Effects of drugs on clinical laboratory tests. *Clin. Chem.*, **18**, 1041.

Young, M. M., Jasani, C., Smith, D. A. and Nordin, B. E. C. (1968), Some effects of ethinyl oestradiol on calcium and phosphorus metabolism in osteoporosis. *Clin. Sci.*, **34**, 411.

Index

Index